Understanding the NEC4 Professional Service Contract

As usage of the NEC (formerly the New Engineering Contract) family of contracts continues to grow worldwide, so does the importance of understanding its clauses and nuances to everyone working in the built environment.

Currently in its fourth edition (NEC4), this set of contracts is different to others in concept as well as format, so users may well find themselves needing a helping hand along the way.

Understanding the NEC4 Professional Service Contract uses plain English to lead the reader through the NEC4 Professional Service Contract's key features and differences from its predecessor, the NEC3 Professional Services Contract, including:

- Main and Secondary Options
- the use of early warnings
- programme provisions
- payment
- compensation events
- preparing tender documents

Common problems experienced when using the Professional Service Contract are signalled to the reader throughout, and the correct way of interpreting each clause explained. The way the contract affects procurement processes, dispute resolution, project management and risk management are all addressed in order to direct the user to best practice.

Written for construction professionals, by a practising international construction contract consultant, this handbook is the most straightforward, balanced and practical guide to the NEC4 Professional Service Contract available. It is an ideal companion for Clients, Contractors, Service Managers, Project Managers, Supervisors, Engineers, Architects, Quantity Surveyors, Subcontractors and anyone else interested in working successfully with the NEC4 Professional Service Contract.

Kelvin Hughes spent 18 years in commercial management with major contractors, then the past 28 years as a consultant, including a four-year senior lectureship at the University of Glamorgan (now University of South Wales). He has been a leading authority on the NEC contract since 1996 was Secretary of the NEC Users' Group for 10 years and has run over 1,700 NEC training courses.

Understanding Construction

Understanding Active Noise Cancellation
Colin H. Hansen (2001)

Understanding the CDM Regulations
Owen Griffiths (2006)

Understanding Building Failures, 4th edition
James Douglas and W. H. Ransom (2013)

Understanding NEC3
Professional Services Contract
Kelvin Hughes (2013)

Understanding NEC3
Engineering and Construction Short Contract
Kelvin Hughes (2014)

Understanding The Building Regulations 6th edition
Simon Polley (2014)

Understanding JCT Standard Building Contracts 10th edition
David Chappell (2017)

Understanding NEC4
Term Service Contract
Kelvin Hughes and Patrick Waterhouse (2018)

Understanding the NEC4 ECC Contract
A Practical Handbook
Kelvin Hughes (2018)

Understanding the NEC4 Professional Service Contract
A Practical Handbook
Kelvin Hughes (2020)

Understanding the NEC4 Professional Service Contract

A Practical Handbook

Kelvin Hughes

Routledge
Taylor & Francis Group
LONDON AND NEW YORK

First published 2021
by Routledge
2 Park Square, Milton Park, Abingdon, Oxon OX14 4RN

and by Routledge
52 Vanderbilt Avenue, New York, NY 10017

Routledge is an imprint of the Taylor & Francis Group, an informa business

British Library Cataloguing-in-Publication Data
A catalogue record for this book is available from the British Library

Library of Congress Cataloging-in-Publication Data
Names: Hughes, Kelvin (Engineering consultant), author.
Title: Understanding the NEC4 professional services contract: a practical handbook / Kelvin Hughes.
Other titles: Understanding the NEC four professional services contract
Description: Abingdon, Oxon; New York: Routledge, 2021. |
Series: Understanding construction | Includes index.
Identifiers: LCCN 2020024817 (print) | LCCN 2020024818 (ebook) |
ISBN 9780367896041 (hardback) | ISBN 9780367896027 (paperback) |
ISBN 9781003020059 (ebook)
Subjects: LCSH: Civil engineering contracts—Great Britain. |
Construction contracts—Great Britain. | NEC Contracts.
Classification: LCC KD1641 .H8346 2021 (print) | LCC KD1641 (ebook) |
DDC 343.4107/8624—dc23
LC record available at https://lccn.loc.gov/2020024817
LC ebook record available at https://lccn.loc.gov/2020024818

ISBN: 978-0-367-89604-1 (hbk)
ISBN: 978-0-367-89602-7 (pbk)
ISBN: 978-1-003-02005-9 (ebk)

Typeset in Times New Roman
by codeMantra

Contents

Preface

I have been involved with the NEC family of contracts since 1995, and in that time, I have advised on numerous projects using the contracts and carried out over 1,700 NEC-based training courses in the UK and abroad.

I was also Secretary of the NEC Users Group from 1996 to 2006 providing support to users of the contracts including seminars and workshops, during which time, I also ran the Users Group Helpline answering queries on the NEC contract from members of the group.

During my involvement with the NEC, I have always felt that there was a need for practical manuals on the various contracts within the NEC family, and their use including worked examples and illustrations, essentially a training course within a book, including, for example, what does an activity schedule look like, how do time risk allowances and float work, how do you assess compensation events, particularly for omissions of work, etc., which are not covered either by the Guidance Notes, or by other text books.

The Guidance Notes that accompany the contract were written by the contract drafters and whilst the Guidance Notes are comprehensive and well-written, I have always felt that they do not adequately act as a practical handbook. Various other books that are available have been written by people who may not be actively involved with NEC contracts and are not training practitioners to use them.

This book is intended to fill the gap where the Guidance Notes leave off and cover issues that may not have been considered or covered within the official publications.

This book covers the NEC4 Professional Service Contract (PSC), which is widely used, but there is very little written about this member of the NEC4 family.

It is intended to be of benefit not only to professionals who are actually using the contract but also to students who need some awareness of the contract as part of their studies.

Whilst a number of practitioners still use the NEC3 Professional Services Contract, the book is primarily aimed at giving guidance to NEC4 Professional Service Contract users, though the structure and contents of the contracts are very similar, and to that end, much of the advice given in this book is of use to all users.

As I have always viewed the NEC contracts as a manual of good practice as well as a contract, particularly in terms of the use of its main and secondary options, risk management through the early warning system, clear requirement for a programme and a disciplined procedure for change management, I have deliberately included within each chapter an overview of the subject area and how other contracts deal with the issues, as well as detailing the NEC4 provisions, and hope that will give the readership a more rounded view of good project management principles. Hopefully, this will also be useful to student readers.

Readers of my previous book, *Understanding the NEC3 PSC Contract – A Practical Handbook*, will see some duplication from that book when reading this book, particularly when giving general advice; this was deliberate, my intention being that each book should stand on its own, rather than one having to read the first book, to properly understand the second and successive books.

Readers may note the absence of case law within the text of the book. This is a deliberate policy on my part, for three reasons.

First, I am not a lawyer, my background being in senior commercial positions within major building contractors, so I felt, and readers may concur, that I was unqualified to quote and to attempt a detailed commentary on any case law.

Second, there has actually been very little case law on the NEC contracts since they were first launched.

Third, and probably the most important, as a contracts consultant with significant overseas experience of all contracts including the NEC, it was always my intention that the book should attract an international readership. NEC was always conceived as an international contract, so including UK case law would probably limit it to a UK readership.

I have included within the text many examples, as I have found when running NEC training courses that delegates understand principles much better if they can be given illustrated and worked examples, and where numbers are involved using real calculations.

Throughout, I have also endeavoured to maintain the NEC drafting policy of using capital letters for defined terms, e.g. "Completion", "Defects" and "Prices", etc. I have also used capital letters for some terms (e.g. "Consultant", "Main and Secondary Options", "Defects Date", etc.), which are in some cases italicised within the contract, to give them prominence within the text.

My personal motivation for writing this book is to collate into one volume a significant quantity of material, which I have gathered over the years, and to share with others a substantial knowledge and experience of the contract.

Note that in January 2019, the publishers of the NEC4 Contracts issued amendments to the contracts. Where appropriate, the amendments to the NEC4 Professional Service Contract and the NEC4 Professional Service Short Contract have been incorporated within this book.

Kelvin Hughes
May 2020

Acknowledgements

As with all my previous books, I would like to extend my continued and sincere thanks to my wife Lesley, who as always gives me the love, the time, the inspiration and the support to fulfil my life's ambitions.

Also, love and thanks to my parents, Maureen and Dennis, who made me who I am, and I hope in return I continue to make them proud.

I would like to dedicate this book to my grandchildren, Grace, Emily, Joshua, Thomas, Meadow and Ellie, who remind me how old I am, but at the same time what it is like to be young and curious!

Finally, I would like to thank Patrick Hetherington of Taylor & Francis Group for his support and unfailing patience during the writing of this book as I juggled writing, running a business and spending time with my family, amid the COVID-19 lockdown.

Introduction

Background to NEC4

The history of the New Engineering Contract (NEC) stems from 1985, when the Council of the Institution of Civil Engineers (ICE) approved a recommendation from its Legal Affairs Committee "to lead a fundamental review of alternative contract strategies for civil engineering design and construction with the objective of identifying the needs for good practice".

There had been a concern for many years that the policy of construction contracts being sector-specific, i.e. building, civil engineering, process, etc., and using a single procurement route was outdated. In addition, there had been concerns for many years about issues such as post-completion prolonged claims negotiations and the time taken to agree final accounts.

In 1986, the specification for the new contract was submitted to the Legal Affairs Committee, and in 1991, the first consultation document was issued with consultation lasting a period of two years. It is perhaps worth mentioning that all of the successive NEC contracts have gone through a consultation process, inviting comments by the NEC Users Group so that potential users can offer comments about the proposed new contracts.

The New Engineering Contract (now NEC4) is now being used substantially within the construction industry and by many Clients in the public and private sectors in several countries.

There are now several national and international flagship projects, and tens of thousands of other projects which have and are being built using the NEC contracts, with most Clients having reported that the contract gives far greater control of time, cost, quality and risk issues, combined with greatly improved relationships between the contracting parties.

In July 1994, Sir Michael Latham's report "Constructing the Team"[1] recommended that the then New Engineering Contract, should be adopted by Clients as the standard form of contract to be used in both the private and public sectors rather than using bespoke forms of contract and suggested that it should become a national standard contract across the whole of engineering and construction work generally. He stated that endlessly refining existing conditions of contract would not solve adversarial problems, but a set of basic principles is required on which modern contracts can be based, and a complete family of interlocking documents is also required.

He also stated that the New Engineering Contract (NEC) fulfilled many of these principles and requirements, but at that time that changes to it would be desirable and the matrix was not yet complete.

Latham also stated that the most effective form of contract should include:

1 A specific duty for all parties to deal fairly with each other and in an atmosphere of mutual co-operation.
2 Firm duties of teamwork, with shared financial motivation to pursue those objectives.
3 A wholly interrelated package of documents suitable for all projects and any procurement route.
4 Easily comprehensible language and with guidance notes attached.
5 Separation of the roles of contract administrator, project or lead manager and adjudicator.
6 A choice of allocation of risks to be decided as appropriate to each project.
7 Where variations occur, they should be priced in advance, with provision for independent adjudication if agreement cannot be reached.
8 Provision for assessing interim payments by methods other than monthly valuation, i.e. mile stones, activity schedules or payment schedules.
9 Clearly setting out the period within which interim payments, failing which there should be an automatic right to compensation, involving payment of interest.
10 Providing for secure trust fund routes of payment.
11 While avoiding conflict, providing for speedy dispute resolution by an impartial adjudicator/referee/expert.
12 Providing for incentives for exceptional performance.
13 Making provision where appropriate for advance mobilisation payments.

The NEC family provides for all of Latham's recommendations.

In June 2005, NEC3, the previous generation of NEC contracts, was launched as a complete review and update of the whole NEC family, including the introduction of two new contracts, the Term Service Contract and the Framework Contract. Since that date, the Supply and Short Supply Contracts have also joined the family.

Alongside the launch of NEC3, the Office of Government Commerce (OGC) announced that after extensive research, they identified NEC3 as the contract that should be adopted by private and public Clients in commissioning their projects.

This was reflected on the inside front cover of the NEC3 contracts when they were launched in June 2005, which stated:

> OGC advises public sector procurers that the form of contract used has to be selected according to the objectives of the project, aiming to satisfy the Achieving Excellence in Construction (AEC) principles. This edition of the NEC (NEC3) complies fully with the AEC principles. OGC recommends the use of NEC3 by public sector construction procurers on their construction projects.

This note was changed in the 2010 reprint to:

> The Construction Clients' Board (formerly Public Sector Construction Clients Forum) recommends that public sector organisations use the NEC4 contracts

when procuring construction. Standardising use of this comprehensive suite of contracts should help to deliver efficiencies across the public sector and promote behaviours in line with the principles of Achieving Excellence in Construction.

It should be stressed that NEC3 is the only contract to have received this recommendation to date.

In June 2017, the NEC4 Contracts were launched and include:

- Engineering and Construction Contract
- Engineering and Construction Subcontract
- Engineering and Construction Short Contract
- Engineering and Construction Short Subcontract
- Professional Service Contract
- Professional Service Subcontract
- Professional Service Short Contract
- Term Service Contract
- Term Service Short Contract
- Design Build and Operate Contract
- Alliance Contract (launched as a consultative edition, then formally in June 2018)
- Supply Contract
- Supply Short Contract
- Framework Contract
- Dispute Resolution Service Contract

Note that, in January 2019, the publishers of the NEC4 Contracts issued amendments to the Contracts. Where appropriate, the amendments to the NEC4 Professional Service Contract and the NEC4 Professional Service Short Contract have been incorporated within the chapters of this book.

Whilst articles written about NEC tend to give prominence to many of the big "flagship" projects carried out using the contracts, of much more significance are the very many more modest size and value projects being procured. It is also in this area that many of the problems have arisen in implementation of the contract, where perhaps less-experienced practitioners may be involved, some of which have hopefully been addressed within this book.

The original objectives of the NEC contracts, and more specifically the Professional Service Contract (PSC), which is the specific topic of this book, were to make improvements under three headings.

Flexibility

The contract should be able to be used:

- for any engineering and/or construction work containing any or all of the traditional disciplines such as civil engineering, building, electrical and mechanical work and also process engineering.

  Previously, contracts had been written for use by specific sectors of the industry, e.g. ICE civil engineering contracts, JCT building contracts, IChemE process contracts and, in turn, consultancy agreements had been written for specific sectors e.g. RIBA Agreements for appointing Architects, RICS Standard Form of Consultant's Appointment for appointing Surveyors, FIDIC Client/Consultant Model Services Agreement for overseas contracts.

- to provide all the normal current options for types of professional service contract such as lump sum, target and cost-reimbursable contracts.

  Previously, professional service contracts were written, primarily as either lump sum or "ad valorem" ("according to value"... of the construction project related to the professional service contracts), so there was no choice of procurement method when using a specific standard form.
- to allocate risks to suit each particular project.

  Previously, contracts were written with risks allocated by the contract drafters, and one had to be an expert in contract drafting to amend the conditions to suit each specific project on which it was used.
- anywhere in the World.

  Previously, contracts included country-specific procedures and legislation and therefore either could not be used or had to be amended to be used in other countries. To that end, FIDIC contracts have been seen as the only forms of contract that could be used in a wide range of countries.

To date, the NEC contracts have been used for projects as widely diverse as airports, sports stadiums, water treatment works, housing projects, in many parts of the world and even research projects in the Antarctic.

Clarity and simplicity

- The contract is written in ordinary language and in the present tense.
- As far as possible, NEC only uses words that are in common use so that it is easily understood, particularly where the user's first language is not English. Previously, obscure words such as "whereinbeforesaid", "hereinafter" and "aforementioned" were commonplace in contracts. It has few sentences that contain more than forty words and uses bullet points to subdivide longer clauses.
- The number of clauses and the amount of text are also less than in most other standard forms of contract, and there is an avoidance of cross-referencing found in more traditional standard forms.
- It is also arranged in a format, which allows the user to gain familiarity with its contents, and required actions are defined precisely thereby reducing the likelihood of disputes.
- Finally, subjective words and phrases such as "fair" and "reasonable" have been used as little as possible as they can lead to ambiguity, so more objective words and statements are used.

Some critics of NEC have commented that the "simple language" is actually a disadvantage as certain clauses may lack definition, and there are certain recognised words that are commonly used in contracts.

There is very little case law in existence with NEC contracts, and whilst adjudications are confidential and unreported, anecdotal evidence suggests that there does not appear to be any more adjudications with NEC contracts than any other, which would tend to suggest that the criticism may be unfounded.

Stimulus to good management

This is perhaps the most important objective of the Professional Service Contract in that every procedure has been designed so that its implementation should contribute to, rather than detract from, the effective management of the work.

In order to be effective in this respect, contracts should motivate the parties to proactively want to manage the outcome of the contract not just to react to situations. NEC intends the parties to be proactive and not reactive. It does also require the parties, and those that represent them, to have the necessary experience (sadly often lacking) and to be properly trained so that they understand how the NEC works.

The philosophy is founded on two principles:

- "Foresight applied collaboratively mitigates problems and shrinks risk"
- "Clear division of function and responsibility helps accountability and motivates people to play their part"

Examples of foresight within the NEC4 Professional Service Contract are the early warning and compensation event procedures, the early warning provision requiring the Client's Service Manager and the Consultant each to notify the other upon becoming aware of any matter which could have an impact on price, time or quality.

A view held by many Clients is that an early warning is something that the Consultant would give and is an early notice of a "claim". This is an erroneous view as first, early warnings should be given by either the Service Manager or the Consultant, whoever becomes aware of it first, the process being designed to allow the Service Manager and the Consultant to share the knowledge of a potential issue before it becomes a problem; and second, early warnings should be notified regardless of whose fault the problem is, it is about raising and resolving the problem and not compensating the affected party.

The compensation event procedure requires the Consultant, within three weeks, to submit a quotation showing the time and cost effect of the event. The Service Manager then responds to the quotation within two weeks, enabling the matter to be properly resolved close to the time of the event rather than many months or even years later.

The programme is also an important management document with the contract clearly prescribing what the Consultant must include within his programme and requiring the Service Manager to "buy into it" by formally accepting (or not accepting) the programme. The programme is then defined as the "Accepted Programme".

In total, the Professional Service Contract is designed to provide a modern method for Clients, Consultants, Contractors, Subcontractors and others to work collaboratively and to achieve their objectives more consistently than has been possible using other traditional forms of contract.

People will be motivated to play their part in collaborative management if it is in their commercial and professional interest to do so.

Uncertainty about what is to be done and the inherent risks can often lead to disputes and confrontation but the Professional Service Contract clearly allocates risks and the collaborative approach will reduce those risks for all the parties so that uncertainty will not arise.

Essential differences between the NEC4 Professional Service Contract and other forms of contract

- The NEC4 Professional Service Contract is part of a family of NEC4 contracts including the Engineering and Construction Contract, Engineering and Construction Subcontract, Engineering and Construction Short Contract, Engineering and Construction Short Subcontract, Professional Service Contract, Professional Service Subcontract, Professional Service Short Contract, Term Service Contract, Term Service Short Contract, Design Build and Operate Contract, Alliance Contract, Supply Contract, Supply Short Contract, Framework Contract and Dispute Resolution Service Contract, allowing all parties, whatever the project or service to be provided, to work under similar conditions.
- *Flexibility* of use: the Professional Service Contract is not sector-specific, in terms of it being used for appointing Consultants on a building, civil engineering, mechanical engineering, process contract, etc. as with other contracts, it can be used for any service on any form of engineering or construction.

  This is particularly useful where a major project such as an airport or a sports stadium can be a combination of building, civil engineering and major mechanical and electrical elements.
- *Flexibility of procurement*: the Professional Service Contract's Main and Secondary Options allow it to be used for almost any procurement method. Again, other contracts do not offer this flexibility.
- *Early Warning*: the Professional Service Contract contains express provisions requiring the Service Manager and the Consultant to notify early warnings, and if required call an "early warning meeting" when either becomes aware of any matter which could affect price, time or quality. Few contracts have this express requirement.
- *Programme*: There is a clear and objective requirement for a detailed programme with method statements and regular updates that provides an essential tool for the parties to manage the project and to notify and manage the effect of any changes, problems, delays, etc. Whilst other contracts contain programme requirements, they do not deal with them in the same detail, and one could imply that they probably do not properly recognise the importance of a programme.
- *Compensation events*: This procedure is unique to NEC and requires the Consultant to price the time and cost effect of a change within three weeks and for the Service Manager to respond within two weeks. There is therefore a "rolling Final Account" with early settlement and no later "end of project" claims for delay and/or disruption. It is also more beneficial for the Consultant in terms of his cash flow as the Consultant is paid *pre agreed* sums rather than reduced "on account" payments, which are subject to later agreement and payment.
- *Disputes*: The contract encourages better relationships, and there is far less tendency for disputes because of its provisions. If a dispute should arise, there are clear procedures as to how to deal with it, i.e. adjudication, tribunal.

The principles of the NEC4 Professional Service Contract (PSC)

The NEC4 Professional Service Contract (the NEC3 version was titled "Professional Services Contract) is used for "the appointment of a supplier of professional services" whether or not an NEC contract is being used for a particular project related to it.

Since its original publication, it has been used for appointing Consultants of all types and disciplines including Engineers, Designers, Quantity Surveyors, Land Surveyors and, of course, Project Managers and Supervisors appointed to act on an Engineering and Construction Contract (ECC).

Under a design and build arrangement, the Contractor would normally appoint the Designer as a Subcontractor using the Engineering and Construction Subcontract or the Professional Service Contract, where the Contractor is the Client of the Consultant or the Professional Service Subcontract.

Whilst the traditional view is that the Professional Service Contract is used by a Client who wishes to appoint a Consultant, the intention is that its use is wider than that, examples to date including a Client appointing a Contractor during the pre-construction phase of a major project to advise on risk and buildability issues, and also a Contractor employing a Consultant as a designer for a design and build project.

The Professional Service Contract can also be used for appointing anyone from any discipline and any location who is providing a professional service, and it has been used, for example, for appointing Consultants to carry out highways and drainage surveys within the UK and to appoint quantity surveying practices in the Middle East.

The only exception would be appointing a Consultant to act as an Adjudicator, where the Professional Service Contract would not be used, as NEC4 already has the "Dispute Resolution Service Contract" (formerly the "Adjudicator's Contract".)

Published contracts for professional services tend to either follow the "all disciplines" approach of the Professional Service Contract e.g. FIDIC Client/Consultant Model Services Agreement (White Book) or separate contracts for separate disciplines such as the Association for Consultancy and Engineering (ACE) Agreements which are as follows:

Agreement 1: Design
Agreement 2: Advisory, Investigatory and other Services
Agreement 3: Design and Construct
Agreement 4: Sub-Consultancy
Agreement 5: Homeowner
Agreement 6: Expert Witness (Sole Practitioner)
Agreement 7: Expert Witness (Firm)
Agreement 8: Adjudicator

There are separate agreements for use in Scotland and also separate schedules for civil and structural engineering and mechanical and electrical engineering for use with Agreements 1 and 3.

The first edition of the Professional Services Contract was published in 1994, the second edition in 1998, the third edition in 2005 and the fourth edition in 2017.

Many consultancy agreements are written in a fairly "loose" form in that the Consultant has priced the services, he invoices the Client and if he feels he is entitled to additional time and or payment, he notifies the Client accordingly.

The Professional Service Contract has been written very much along the lines of a contract between a Client and a Contractor in that it has a range of payment options similar to a construction contract, requires a detailed programme from the Consultant and the Consultant pays delay damages if the services are late. The Consultant is essentially more accountable for his actions and for the deliverables required by the contract he has with the Client.

NEC4 Professional Service Short Contract

The NEC4 Professional Service Short Contract is used for the appointment of a Consultant where the professional service does not require sophisticated management techniques, is for straightforward work and imposes only low risks upon both Parties.

The NEC4 Professional Service Short Contract is discussed in detail in Chapter 12 of this book.

Arrangement of the Professional Service Contract

The Professional Service Contract includes the following sections:

- Core clauses

  1 General
  2 The Consultant's main responsibilities
  3 Time
  4 Quality Management
  5 Payment
  6 Compensation events
  7 Rights to material
  8 Liabilities and insurance
  9 Termination

- Main option clauses

  Option A: Priced contract with activity schedule
  Option C: Target contract
  Option E: Cost reimbursable contract (formerly referred to in the NEC3 Professional Services Contract as a "time based contract")

  Note the absence within the NEC4 Professional Service Contract of "Option G – Term Contract" which was included within the NEC3 Professional Services Contract, but Clients should now use the Term Service Contract if they need a term service option for a consultancy service, the Consultant then being "the Contractor".

  A question that the author has been asked many times is "why are the Main Options A, C, and E, rather than A, B and C"? The simple answer is that Option A under all of the NEC Contracts is always a "priced contract with activity schedule", Option C is always a "target contract" and Option E is always a "cost reimbursable contract".

- Dispute resolution

  Option W1: Dispute resolution procedure (used unless the Housing Grants, Construction and Regeneration Act 1996 applies)
  Option W2: Dispute resolution procedure (used in the United Kingdom when the Housing Grants, Construction and Regeneration Act 1996 applies)

- Secondary option clauses

Option X1:	Price adjustment for inflation
Option X2:	Changes in the law
Option X3:	Multiple currencies
Option X4:	Ultimate holding company guarantee
Option X5:	Sectional Completion
Option X6:	Bonus for early Completion
Option X7:	Delay damages
Option X8:	Undertakings to Others
Option X9:	Transfer of rights
Option X10:	Information Modelling
Option X11:	Termination by the Client
Option X12:	Multiparty collaboration
Option X13:	Performance Bond
Option X18:	Limitation of Liability
Option X20:	Key Performance Indicators
Option Y(UK)1:	Project Bank Account
Option Y(UK)2:	Housing Grants, Construction & Regeneration Act 1996
Option Y(UK)3:	The Contracts (Rights of Third Parties) Act 1999
Option Z:	Additional conditions of contract

NB Options X14–X17 and X19 are not used, but they feature in other NEC contracts.

- Contract Data

Other documents to be used with the Professional Service Contract include

- the Scope
- the Accepted Programme
- an activity schedule (Main Option A or C)
- other documents resulting from choosing various secondary options e.g. ultimate holding company guarantee (Option X4) and performance bond (X13)

The Main and Secondary Options

The Main Options

The three Main Options A, C and E enable Clients to select a procurement strategy and payment mechanism most appropriate to the project and the various risks involved.

Chapter 5 includes more details on payment mechanisms and the contractual requirements under each Main Option. Essentially, the main options differ in the way the Consultant is paid.

Whilst many traditional contracts are based on bills of quantities, or in the case of Consultants schedules of work, there has been a movement away from the use of traditional bills and towards payment arrangements such as milestone payments and activity schedules, with payment based on progress achieved, rather than quantity of work done.

There is also an increasing use of target cost contracts which has been encouraged by the increasing use of partnering arrangements, the better sharing of risk and also the continued growth of NEC contracts which provide for target options.

The Main Options principles

Option A: Priced contract with activity schedule
Option C: Target contract
Option E: Cost-reimbursable contract

- Option A is a lump sum-priced contract in which the risks of being able to carry out the service at the agreed prices are largely borne by the Consultant.
- Option C is a target contract in which the Client and the Consultant share the financial risks in pre-agreed proportions.
- Option E is a cost-reimbursable contract in which the financial risks of being able to carry out the work are largely borne by the Client.

There is no option for the Consultant to apply a fee percentage against the values of the construction project, and this is because the value of the services and any changes to that value do not necessarily or fairly reflect upon the input of the Consultant. Such an arrangement would also not motivate the Consultant to consider economies or value-engineered solutions in respect of the price of the project.

Option A: Priced contract with activity schedule

Option A is a lump sum-priced contract and normally used where the Client is able to specify accurately what he requires. This includes a clear statement of what is required, e.g. design brief, performance specification and a statement of the purpose of the project.

An activity schedule is a list of activities prepared by the Consultant which he expects to carry out in Providing the Service. When the Consultant has priced it, the lump sum for each activity is the Price to be paid by the Client upon completion of that specific activity.

Option A is therefore effectively a stage payment contract, and as payment is linked to completion of activities, the Consultant must plan and carry out his work effectively with the cash flow requirements for both parties being clearly visible.

An activity is a part of the whole services as defined in the Scope.

Establishing the quantities of work involved to achieve the completion of each activity is the responsibility of the Consultant. The price for each activity is in effect a lump sum for that activity and must include for everything necessary to complete the activity. The sum of the tendered lump sums for each of the activities is the tendered Price for the whole of the services.

Activity schedules are activities in the programme for which the Consultant gets paid when he has completed that activity. The activity schedule is therefore closely linked to the construction programme prepared by the Consultant. It is therefore sensible to arrange that the Consultant also prepares the detail and breakdown of the activity schedule.

Theoretical advantages, which the use of activity schedules should give and which appear to be materialising in practice:

- if programme and activity schedule preparation are integrated activities, it should lead to a more comprehensive tender.
- payment is linked to completion of an activity or group of activities, and the cash flow requirements for both parties are therefore more visible.
- in order to receive payment as planned, the Consultant has to complete an activity by the assessment date. Consequently, he has to programme realistically and is motivated to keep to that programme during construction.
- Consultants are not paid for changes in quantity of the permanent work, unless an instruction changes the Scope. This transfers some risk to the Consultant.
- the assessment of the effects of compensation events is easier with activity schedules. Any change in resources or methods associated with an activity can be compared with those stated in the Accepted Programme before the compensation event occurred.
- with an activity schedule, the assessment of the payments due to the Consultant is easier and involves significantly less time than with the bill of quantities.

The possible disadvantages are:

- the tendering Consultant will spend additional time at tender stage establishing the quantity of actual work for costing and including in the tender activity/lump sum.
- the risk of the accuracy of the quantities is transferred to the Consultant.
- the Client must be extremely diligent in setting down in the Scope the necessary information describing exactly what the Consultant is required to do in Providing the Services.

Establishing and taking off the quantities of work involved to achieve the completion of each activity is the responsibility of the tendering Consultants. The price for each activity is in effect a lump sum for that activity and must include for everything necessary to complete the activity. The sum of the tendered lump sums for each of the activities is the tendered prices (the total of the Prices) for the whole of the services.

Under Option A, the Consultant prepares and submits an activity schedule with his tender. The total of the Prices in the activity schedule is then the Consultant's price for Providing the Service. The contract requires that the activities on the Activity Schedule relate to operations on each programme.

It is important to recognise that the Activity Schedule has two primary functions:

- it shows how the tendering Consultants have built up their price and can therefore be used as part of the tender assessment process.
- it is used to calculate the Price for Work Done to Date

The price for each activity is in effect a lump sum for carrying out and completing that activity and must therefore include everything necessary to complete that activity.

The Client does not provide a bill of quantities with Option A, the taking off of the quantities of work involved to achieve the completion of each activity and the pricing the activity being the responsibility of the tendering Consultants.

There are certain advantages to the Client using Option A:

- there is no requirement for a bill of quantities to be prepared for tendering Consultants to price. This will save time and money at the pre-tender stage.
- the Consultant holds the risk of inaccuracies in the quantities or missing an item of work stated within the Scope. However, it must be reinforced that the Client holds the risk of inaccuracies and omissions within the Scope.
- the assessment of the Price for Work Done to Date is easier and quicker than with the other main options. There is no requirement to remeasure the work and apply quantities to rates and prices to calculate the amount due to the Consultant.
- programme and activity schedule preparation are linked as integrated activities which would tend to lead to a more comprehensive tender.
- payment is linked to completion of an activity or group of activities related back to the programme, so cash flow requirements for both parties are more visible.
- in order to receive payment, the Consultant has to complete an activity by the assessment date, so he has to price and programme realistically and is motivated to keep to that programme during construction.
- the assessment of the effects of compensation events is related back to the activity schedules. Any change in resources or methods associated with an activity can be compared with those stated in the Accepted Programme before the compensation event occurred.

The possible disadvantages to the Client are:

- as there is no specific document that all the tendering Consultants price, it is difficult to assess tenders on a "like for like" basis. Some Clients cite that as an advantage, saying that they do not require the facility to check tenders on a line-by-line basis anyway

It is not uncommon to assist the assessment of tenders for Clients to issue templates for the Consultant to price covering the main construction elements as part of the tender with the tendering Consultants then inserting the activities under within each element. In that way, each element will have been priced on a "like for like" basis, but the activities within each element may have been described and priced differently by each tenderer.

If a template is used, Clients should be wary of them being too prescriptive and almost being seen by tenderers as a bill of quantities.

- the tendering Consultants have to spend time, and money, in taking off the quantities of work in order to establish and populate the activity schedule. In order to reduce the tenderers efforts in producing their own quantities, some Clients issue quantities to tenders, but these must be stressed as for indicative purposes only, and tenderers should not rely on those quantities.
- the Client holds the risk of clearly defining the Scope in order that the Consultant can prepare the Activity Schedule.

There may be an activity schedule which is a list of activities prepared by the Consultant which he expects to carry out in providing the services. When the Consultant has priced it, the lump sum for each activity is the price to be paid by the Client in the assessment following the completion of that activity.

When activity schedules are used, payment is linked to completion of activities, the Consultant must plan and carry out his work effectively with the cash-flow requirements for both parties being clearly visible. Administration of payments is therefore fairly simple.

An example of an activity schedule for an Architect could be based on elements of the RIBA Outline Plan of Work, each stage of which should be defined by deliverables in the form of submitted drawings, a report or some other document so that whether an activity is completed or not can be assessed without any doubt.

Stage A – Appraisal

- Identification of Client's needs and objectives, business case and possible development constraints.
- Preparation of feasibility studies and assessment of options to enable the Client to decide whether to proceed.

Stage B – Design Brief

- Development of initial statement into Design Brief, confirming key requirements and constraints. Identification of procurement method, procedures, organisational structure and range of Consultants to be engaged.

Stage C – Concept

- Implementation of Design Brief and preparation of additional data.
- Preparation of Concept Design including outline proposals for structural and building services outline specification and preliminary cost plan.
- Review of procurement route.

Stage D – Design Development

- Development of concept design to include structural and building services, updated outline specifications and cost plan.
- Completion of Project Brief.
- Application for detailed planning permission.

Stage E – Technical Design

- Preparation of technical designs and specifications, sufficient to co-ordinate components and elements of the project and information for statutory standards and construction safety.

Stage F – Production Information

- Preparation of production information in sufficient detail to enable a tender or tenders to be obtained.
- Application for statutory approvals.
- Preparation of further information for constriction required under the building contract.

Stage G – Tender Documentation

- Preparation and/or collation of tender documentation in sufficient detail to enable a tender or tenders to be obtained.

Stage H – Tender Action

- Identification and evaluation of potential Consultants and/or specialists for the project.
- Obtaining and appraising tenders; submission of recommendations to the Client.

Stage J – Mobilisation

- Letting the building contract, appointing the Consultant.
- Issuing of information to the Consultant.
- Arranging site hand over.

Stage K – Construction to Practical Completion

- Administration of the building contract to Completion.
- Provision to the Consultant of further information as and when reasonably required.
- Review of information provided by the Consultants and specialists.

Stage L – Post-Practical Completion

- Administration of the building contract after Practical Completion and making final inspections.
- Assisting building user during initial occupation period.
- Review of project performance in use.

Option C – Target contract

Target contracts are a development of cost-reimbursable contracts and are used where the extent of work to be done is not fully defined or where risks are greater.

Such financial risks are shared, not necessarily equally, between the Client and the Consultant.

The Consultant tenders a target price using an activity schedule which he prepares.

During the course of the contract, the Consultant is paid the Price for Service Provided to Date in accordance with Defined Cost plus Fee, and at the end of the contract, the Consultant is paid (or pays) his share of the difference between the final total of the Prices and the final Price for Service Provided to Date according to a formula stated in the Contract Data. If the final Price for Services Provided to Date is greater than the final total of the Prices, the Consultant pays his share of the difference.

Option C is a target contract based on an activity schedule, again as with Option A normally prepared by the Consultant. It must be emphasised that the activity schedule is not used for payments in the same way as with an Option A contract. In Option C, the activity schedule acts as a guide when assessing tenders, but payments are based on Defined Cost incurred, not completion of activities.

Target contracts are a development of cost-reimbursable contracts and are a combination of a lump sum price and open book cost-reimbursable payments.

The advantages of target contracts are that the Parties initially have certainty of price, with the Consultant being incentivised to make cost savings for the benefit of the Client and himself.

This option is normally used where the Client knows what he wants and is able to clearly define it through the Scope, so that the Consultant can price and prepare the activity schedule, but he sees a benefit in sharing risk and opportunity with the Consultant thereby encouraging collaboration.

Financial risks are shared proportionally between the Client and the Consultant through the Consultant's share percentages. Many wrongly call Option C, the "Partnering Option" although it does lend itself to the principles of partnering, collaborative working and associated risk/opportunity sharing.

Coincidentally, there is a Secondary Option X12 that was formerly called "Partnering", which is now called "Multiparty Collaboration", which may have given rise to the confusion. Option X12 is described in detail later within this chapter.

The Consultant tenders a price and includes an activity schedule in the same way as he would under an Option A contract.

This price, when accepted, is then referred to as the "target". The Consultant also tenders his percentage(s) for Fee.

The original target is referred to as the Total of the Prices at the Contract Date.

- The target price includes the Consultant's estimate of Defined Cost plus other costs, overheads and profit to be covered by his Fee.
- The Consultant tenders his Fee in terms of percentages to be applied to Defined Cost.
  - During the course of the contract, the Consultant is paid Defined Cost plus the Fee.
  - The target is adjusted for compensation events and also for inflation (if Option X1 is used)
- On Completion, the Client assesses the Consultant's share in accordance with Clause 54, which, it has to be said, is at best a confusing clause, though the Guidance Notes clarify the clause. The Consultant then pays or is paid his share of the difference between the final total of the Prices and the final Payment for Work Done to Date according to a formula stated in the Contract Data. This motivates the Consultant to decrease the costs. Many NEC practitioners refer to this sharing of risk and opportunity as "pain and gain".
- It often comes as a surprise to NEC users that the terms "pain and gain" do not appear anywhere within the NEC contracts, nor do the terms "target price" or "target cost".

Under Options C and E, the Consultant is required to submit forecasts at the intervals stated in the Contract Data of the total Defined Cost and expenses, which advises the Client of his potential outturn cost.

Option E – Cost reimbursable contract

Option E is a cost-reimbursable contract (formerly referred to in the NEC3 PSC as a "time based contract"), which is normally used where the definition of the service to be provided is inadequate for a lump sum price or even as a basis for a target price and yet an early start is required.

In such circumstances, the Consultant cannot be expected to take risks. He carries minimum risk and is reimbursed his Defined Cost plus Fee.

A criticism of cost-reimbursable contracts such as Option E is that it gives little incentive for the Consultant to minimise the costs during the services, but this particular strategy may be appropriate where time or quality are overriding priorities or where the scope is not known.

For instance:

- where the Scope is uncertain e.g. refurbishment projects,
- where extreme flexibility is required e.g. for enabling work,
- where a high level of Client involvement is envisaged and
- in emergency work, which would be time-driven.

The option allows development of the design as the services proceed and permits maximum flexibility in allocation of design responsibility.

The Secondary Options

Option X1: Price adjustment for inflation (used only with Options A and C)

This option can only be used with Options A and C. It is particularly suitable for service contracts where inflation is unpredictable, or where the period of the service is lengthy, say more than two years and thus where inflation may represent a high risk.

It also has to be said that when using Option X1, the Clause itself and the actual calculations can be extremely complicated.

The Client should make the decision at the time of preparing the tender documents as to whether inflation for the duration of the contract is to be:

- The Consultant's risk: in which case, he should not select Option X1
- The Client's risk: in which case, he should select Option X1

The default within the Professional Service Contract is that the contract is "fixed price" in terms of inflation, i.e. the Consultant has priced the service to include any inflation he may encounter during the period of that service.

If Option X1 is chosen, the Prices are adjusted for inflation as the work progresses, by means of a formula. Inclusion of this Option effectively transfers the risk of inflation to the Client.

The Option uses the formula method utilising published indices for rises in costs for various elements of the service.

- When used with Option A, the formula is applied to the amount due to the Consultant. The Price Adjustment Factor is calculated each time an amount becomes due – not only at the anniversary date.
- When used with Main Option C, the amount for price adjustment is applied only to the Prices (the target).

Note that under Option C, whether Option X1 is selected or not, the Price for Work Done to Date is the current cost at the time is incurred. Option X1 is then added to the Total of the Prices (the target).

The key components of the formula are:

- the "Base Date Index" (B) used for the first inflation adjustment date is the latest index available before the Base Date,
- the "Latest Index" (L) is the latest available index before the date of an assessment of an amount due and
- the "Price Adjustment Factor" is the total of the products of each of the proportions stated in the Contract Data multiplied by (L – B)/B for the index linked to it.

Option A

Under Option A, the amount due includes an amount for price adjustment which is the sum of:

- the change in the Price for Service Provided to Date since the last assessment of the amount due multiplied by the PAF and
- the amount for price adjustment included in the previous amount due.

As an example, for simplicity ignoring correcting amounts:

The change in the Price for Service Provided to Date = £150,000
The Base Date Index (B) = 280.0
The Latest Index (L) = 295.5
The Price Adjustment Factor is therefore (L – B)/B
= (295.4 – 280.0)/280.0
= 0.055
Inflation since the base date is therefore 5.5%
The amount due is therefore £150,000 × 0.055 = £8,250.00

Option C

Under Option C, each time, the amount due is assessed (normally monthly), an amount for price adjustment is added to the total of the Prices, i.e. the target.

Note that the compensation events are priced using the Defined Cost at base-date level for amounts calculated from rates in the Contract Data and the Defined Cost current at the dividing date used in assessing the compensation event adjusted to the base date by dividing by one plus the PAF for the last assessment of the amount due before that dividing date, for other amounts.

Option X2: Changes in the law

This option transfers the financial risk of changes in the law after the Contract Date from the Consultant to the Client. This risk may be high in some countries or where there is a long period of service.

As with Option X1, the default is that the contract is "fixed price" in terms of changes in the law, i.e. the Consultant has priced the work to include any changes in the law he may encounter during the entire period of the contract.

If Option X2 is chosen, and a change in the law occurs after the Contract Date, the Service Manager notifies the Contractor of a compensation event. The Prices may be increased or reduced.

Option X3: Multiple currencies (used only with Option A)

The currency of the contract is stated in Contract Data Part 1. However, Option X3 provides for items or activities to be paid or partly paid in an alternative currency, the items or activities, the currency and the total maximum payment in this currency to be listed in the Contract Data, beyond which payments are made in the currency of the contract.

The exchange rates, their source and date of publication are also referred to in the Contract Data.

Option X4: Ultimate holding company guarantee (formerly under the NEC3 Professional Services Contract "parent company guarantee")

This form of guarantee is given by a parent company (or ultimate holding company) to guarantee the proper performance of a contract by one of its subsidiaries (the Consultant), who, whilst having limited financial resources himself, may be owned by a larger financially sound parent company.

In most cases, it is the ultimate holding company that provides the guarantee, but sometimes, particularly when the ultimate holding company is in another country, the company may just be a company further up the chain within the group, perhaps the national parent who has sufficient assets to provide the required guarantee.

If an ultimate holding company guarantee is required, it may either be provided by the Contract Date or within four weeks of the Contract Date.

The ultimate holding company guarantee should have an expiry date, which could be completion of the project, the Defects Date or may even include the six, ten or twelve-year limitation period following completion of the services to cover any liability for potential latent Defects, the expiry date being defined within the Scope.

Ultimate holding company guarantees are normally used as an alternative to a performance bond (Option X13).

Such a guarantee is cheaper than a performance bond, as the Consultant will normally just charge an administration fee rather than the case of performance bond, where the Consultant is actually paying a premium for an insurance policy, but it may give less certainty of redress because it is not supplied by an independent third party, so it is dependent on the survival, and the ability to pay, of the parent/ultimate holding company.

However, whilst accepting less independence, ultimately, holding company guarantees for the proper performance of the contract can be more advantageous than bonds.

Rather than receiving a fixed amount in compensation, the parent company may be obliged to either complete the service in accordance with the Consultant's original obligations on behalf of its subsidiary company or fund the completion of the contract by others; so, effectively, the completion is guaranteed.

In addition, further recompense can be sought for time delays in completion through the normal clauses incorporated in the contract. In the case of performance bonds, the "guarantee" is that a sum of money is available to at least partly compensate the Client in the event of the Consultant's default.

Because the financial strength of the ultimate holding company may be linked to that of the Consultant, such a guarantee will be acceptable only if the ultimate holding company is financially strong enough, and its financial resources are largely independent of those of the Consultant. Obviously, if the insolvency is not limited to the Consultant but also relates to the parent group, an ultimate holding company guarantee will be virtually useless, save for any entitlement under insolvency law.

Option X5: Sectional Completion

If the Client requires sections of the service to be completed by the Consultant before the whole of the service is completed, then Option X5 should be chosen.

References in the contract to the service, Completion and Completion Date will then apply to either the whole of the service or a specified section.

Contract Data Part 1 should give a description of each section and the date by which it is to be completed.

Option X5 may be selected together with Option X6 (bonus for early Completion) and/or Option X7 (delay damages).

Option X6: Bonus for early Completion

Often, early completion of the service would be of benefit to the Client, for example, the Consultant is designing a stadium, and early appointment of the Contractor and completion of construction would allow a sporting event or concert to take place.

Option X6 provides for the Client to pay a bonus to the Consultant for early completion of the service. The amount of the bonus is stated by the Client on a "per day" basis in Contract Data Part 1 and is calculated from the earlier of Completion and the Completion Date.

Option X7: Delay damages

Delay damages in the Professional Service Contract are normally referred to in other contracts as "liquidated damages".

Delay damages are pre-defined amounts inserted into Contract Data Part 1 and paid or withheld from the Consultant in the event that he fails to complete the service by the Completion Date.

The amount included within the contract for delay damages should be a "genuine pre estimate of likely losses" and not a penalty.

Many contracts require the Contract Administrator to issue some form of certificate confirming that a party failed to complete on time and also for the Client to notify that party in writing that he will be withholding the relevant damages, but the Professional Service Contract provides for the Consultant to pay delay damages at the rate stated in the Contract Data until Completion or the date on which the Client has taken over the service, whichever is earlier, without having to certify that the Consultant has defaulted.

Also, many contracts require the Contract Administrator to value the services, and the Client then deducts the liquidated damages from the amount otherwise due to the Consultant, but the Professional Service Contract requires the Service Manager to deduct amounts to be paid or retained from the Consultant within his assessment and certificate.

If the Completion Date has been changed to a later date by a compensation event after delay damages have been paid by the Consultant, the damages are repaid back to the Consultant with interest.

Option X8: Undertakings to Others (formerly under the NEC3 Professional Services Contract "collateral warranty agreements")

The NEC4 contracts do not generally provide for collateral warranties or standard forms for their use, though Secondary Option X8 provides for the Consultant to enter into a collateral warranty agreement which has been renamed "undertakings to Others".

A collateral warranty or undertaking to Others is an agreement associated with another contract, collateral meaning "additional but subordinate" or "running side by side", the collateral warranty being entered into by a party to the primary contract.

These are different to traditional contracts, in that the two parties to the warranty do not have any direct contractual relationship with each other.

They are useful for creating ties between third parties, but the main disadvantage is the expense of providing them, particularly where there are many interested third parties such as with housing and commercial properties. When one also considers that parties can change their names or business status during the life of a warranty, then even a medium-sized project can involve 30 or more warranties.

Different parties will also have different requirements for such warranties; therefore, the precise terms of the agreement need to be considered in each case carefully so that they are drafted to provide the required protection.

The basic ingredients of a collateral warranty or undertaking to Others are usually:

i Definitions

There will be a brief description of the services and the warranty/undertaking.

The "contract" is the contract to which the warranty/undertaking is collateral, and the date should be specified. Since the obligations of the Consultant to the Beneficiary parallel the obligations of the Consultant to the Client under the contract for the services, the Beneficiary should always ask to see a copy of the contract. If, for example, there are restrictions on the Consultant's liability in the principal contract that might affect the rights of the Beneficiary under the collateral warranty.

A copy of the principal contract may be attached to the collateral warranty agreement.

ii The warranty

The Consultant gives warranties/undertakings to the Beneficiary that he has exercised and will continue to exercise reasonable skill and care in his obligations to the Client under the principal contract including compliance with any design obligations, performance requirements and correction of Defects.

The warranty should contain a statement that the Warrantor shall have no greater liability to the Beneficiary than they would have under the primary contract.

The Consultant also does not have any liability to the Beneficiary if he fails to complete the services under the contract on time, as that will usually be covered by delay damages under the principal contract.

There should be a statement that the Warrantor shall have no liability under warranty in any proceedings commenced more than twelve years after the date of Completion of the services.

iii Naming of the parties

The named parties can be, for example, the Consultant and the end user, a Subconsultant and the Client or a Consultant and the Client.

The party that gives the warranty is known as the Warrantor; the other party is known as the Beneficiary.

iv Prohibited Materials

The Contractor may be required to give a warranty that he will not specify or use Materials which are known to be prohibited or deleterious to health and safety.

In addition, the use of deleterious Materials may contravene relevant legislation or regulations as well as being expressly prohibited in the contract between the Developer and the Contractor.

v Copyright

Normally, the Beneficiary under a collateral warranty agreement will have the right to use designs and other documents prepared by the Consultant but only in connection with the project for which those design documents are prepared, copyright to these documents remaining with the Consultant.

vi Insurances

There is normally a requirement for the Warrantor to maintain insurances including Professional Indemnity (PI) insurance where applicable for a stated minimum level of cover and for a period of twelve years after Completion of the Services.

vii Assignment

There is normally a provision allowing the Beneficiary up to two assignments of the warranty.

viii Step-in rights

Step-in rights can allow a funder to take over the Client's role and become the Client under the primary contract. This is particularly important where the Client is unable to pay the Consultant, the Consultant having to notify the Beneficiary before he terminates the contract, thus allowing the Beneficiary to "step in" and take the Client's place, with an obligation to pay the Consultant any money that is outstanding. The Beneficiary does not have any obligation to step in, only the right to do so. As the effect of this clause is to vary the principal contract, it is important that, when the collateral warranty contains such a clause, the Client should be a party to that document.

ix Law and jurisdiction

It is important to specify the law of the contract and the warranty and the jurisdiction of the courts that will have jurisdiction.

Such warranties are a relatively recent feature of contracts, coming into use in the past twenty-five years following some notable court cases in which it was judged that it was not possible to recover damages for negligence in relation to defects in construction projects as it was held to be economic loss, which is not recoverable in tort.

Consequently, such claims had to be brought as claims for breach of contract, where economic loss is recoverable. It therefore became necessary to establish a contract between parties involved in a construction project, such as Designers or Subcontractors and parties such as funders, present and future purchasers, who are not parties to the primary contracts but who have an interest in the project, hence the "collateral warranty".

Different parties will also have different requirements for collateral warranties; therefore, the precise terms of the collateral warranty agreement need to be considered in each case carefully so that they are drafted to provide the required protection.

Collateral warranties are executed as a "deed" also referred to as "under seal", which differs from a simple contract in two respects:

- First, in English law, the time after a breach of contract has occurred within which one party can sue another is twelve years, whereas the time for a simple contract executed "under hand" is six years.

- Second, whereas under English law, consideration is needed for a contract to be effective, and this is not the case with a deed. For a company to execute a deed effectively, the document must be signed by two directors or a director and the company secretary, unless the Articles of Association contain some other requirement.

Option X9: Transfer of rights

Under Clause 70.1, the Client has the right to use material supplied by the Consultant for the purpose stated in the Scope, but the Consultant retains ownership and copyright to anything he produces including design and all associated documents.

Under Option X9, the Client owns the rights over materials prepared for the contract except as stated otherwise in the Scope. This effectively reverses Clause 70.1, the Consultant obtaining other rights for the Client, including equivalent rights from Subcontractors, the Consultant providing the relevant documents which transfer these rights to the Client.

Option X10: Information modelling

Note that Option 10 was titled "Employer's Agent", but the NEC4 PSC now has a "Service Manager" who acts for the Client

Option X10 is now a new Option which provides for Information Modelling. Otherwise known as "Building Information Modelling (BIM)".

Some defined terms should be considered in reviewing Option X10.

- The Information Plan is submitted by the Consultant to the Service Manager who then accepts or does not accept within two weeks of receiving it. The Information Execution Plan is then defined as the *information execution plan* (identified within the Contract Data) or the latest submitted by the Contractor and accepted by the Service Manager.
- Project Information is provided by the Consultant and used to create or change the Information Model.
- The Information Model is the electronic integration of the Project Information and other information provided by the Client and other Information Providers. The Client is liable for any fault or error in the Information Model, unless there is a Defect in the Project Information provided by the Contractor. The Contactor is required to provide insurance against claims made in respect of failure to provide the Project Information using reasonable skill and care.
- The Information Model Requirements are the requirements identified in the Scope for creating or changing the Information Model.

There is also provision within the Option for early warnings where something could affect the Information Model, and for the Contractor to include within his quotation for a compensation event where the Information Execution Plan is altered by a compensation event.

Option X11: Termination by the Client

This option provides for the Client to terminate the Contractor's employment for a reason not stated in the Termination Table by notifying the Service Manager and Contractor.

If the Client does terminate, in respect of payments, the Consultant is entitled to payments as if he himself had terminated:

Amount A1

- An amount due as for normal payments.
- Other Defined Cost in reasonable expectation of completing the whole of the service.
- Any amounts retained by the Client.

Amount A3

- For Options A and C, the fee percentage applied to the difference between the original total of the Prices and the Price for Service Provided to Date.
- For Option E, the fee percentage applied to the difference between the first forecast of the Defined Cost for the service and the original total of the Prices and the Price for Service Provided to Date less the Fee.

Option X12: multi-party collaboration (not used with Option X20) (formerly under the NEC3 Professional Services Contract "Partnering")

This option enables a multi-party partnering agreement to be implemented. In this case, Option X12 is used as a Secondary Option common to the contract which each party has with the body which is paying for the work. The parties together make up the partnering team.

The content of Option X12 is derived from the "Guide to Project Team Partnering" published by the Construction Industry Council (CIC). It is estimated that Option X12 is used on 9% of NEC4 contracts (RICS Contract in Use Survey 2007).

It must be stressed that no legal entity is created between the Partners, so it is not a "partnership" as such.

Some definitions need to be explained:

i The Partners are those named in the Schedule of Partners.
ii An Own Contract is a contract between two Partners which includes Option X12.
iii The Core Group comprises the Partners listed in the Schedule of Core Group Members.
iv Partnering Information is information that specifies how the Partners work together.
v A Key Performance Indicator (KPI) is an aspect of performance for which a target is attached in the Schedule of Partners.

Each Partner, represented by a single individual, is required to work with the other Partners in accordance with the Partnering Information to achieve the Client's objective stated in the Contract Data and the objectives of every other Partner.

The Core Group, led by the Promoter's Representative, its members selected by the Partners, acts and takes decisions on behalf of the Partners. The Core Group also keeps up to date, a Schedule of Core Group Members and a Schedule of Partners.

The Partners are required to work together, using common information systems, and a Partner may ask another Partner to provide information which he needs to carry out the work in his Own Contract. Each Partner gives an early warning to the other Partners when he becomes aware of anything that could affect another Partner's objectives.

The Core Group may give an instruction to the Partners to change the Partnering Information, which is a compensation event.

The Core Group also maintains a timetable showing the Partners' contributions. If the Consultant needs to change his programme to comply with the timetable, then it is a compensation event.

Each Partner also gives advice, information and opinion to the Core Group where required.

Each Partner must also notify the Core Group before subcontracting any work, though it does not say that the Core Group is required to respond to the notification.

Finally, Option X12 provides for KPI with amounts paid as stated in the Schedule of Partners. The Client may add a KPI to the Schedule of Partners but cannot delete or reduce a payment.

Option X13: Performance bond

A performance bond is an arrangement whereby the performance of a contracting party (the Principal) is backed by a third party (the Surety), which could be a bank, insurance company or other financial institution that should the Principal fail in his obligations under the contract, normally due to the Principal's insolvency, the Surety will pay a pre-agreed sum of money to the other contracting party (the Beneficiary). Whilst bonds are not always linked to insolvency, contracts normally give remedies in the event of default, e.g. defects provisions, retention, withholding of payment, delay damages, etc. It is where the defaulter is not able to provide a remedy as he is insolvent that bonds normally provide the remedy.

If a performance bond is required from the Consultant, it should be provided by the Contract Date. If it is not provided by the Contract Date, it must be provided within four weeks of the Contract Date.

The amount of the bond, usually 10% of the contract value, must be stated in Contract Data Part 1, and the form of the bond may be stated in the Scope.

In the event of the default by the Consultant, the Client is not paid the full amount of the bond, but the amount of the bond is the maximum amount available to meet the Client's costs incurred as a direct result of the default.

Normally, the value of the performance bond does not reduce, but they should have an expiry date, which could be Completion of the project, the Defects Date or may even include the six, ten or twelve-year limitation period following Completion of the service to cover any liability for potential latent defects, the expiry date being defined within the Scope.

It is worth mentioning that payment of a performance bond is often very difficult and subject to a number of conditions, for example, clear evidence of the Consultant's inability to perform his obligations and also unequivocal evidence of the Client's loss as a direct result of the Consultant's default, which in truth may not be fully known for some time after the Consultant's insolvency.

Clearly, the expiry date for the bond has to be kept in mind if it takes some time to ascertain the Client's actual loss as the right to claim may have expired.

A performance bond may be provided by a bank, an insurance company or a specialist bond provider. Clearly, if a bank provides a bond, there may be some implications with regard to further credit as a safeguard should the bond be called, the bank also pursuing any amount paid from their customer, whereas, with an insurance company, the risk is accepted through the policy.

The bank or insurer which provides the performance bond must be accepted by the Client. It is important to note the difference between an insurance policy and a bond in that an insurance policy is a contract between *two* parties, the insured and the insurer, the insurer guaranteeing that he will pay a third party, who may not be specifically named, should an event occur, whilst a bond is a contract between *three* parties, the Contractor, the Client and the Surety, all of whom are named within the agreement, no other party being able to benefit from the bond.

The cost of the performance bond will normally be governed by:

- the technical ability of the Consultant;
- the usual type of work undertaken by the Consultant;
- the size of the project;
- contracts already bonded for the Consultant;
- the overall management of the Consultant's business and
- the financial stability of the Consultant.

Option X18: Limitation of liability

If the Consultant causes any loss or damage to the Client or to his property, the Consultant would normally be liable for the full cost of the remedy/loss; however, Clause X18 provides for the liability for indirect or consequential loss or latent Defects.

The Consultant is not liable unless details are notified to the Consultant before the end of liability date.

Option X20: Key Performance Indicators (not used with Option X12)

Performance of the Consultant can be monitored and measured against Key Performance Indicators (KPIs using Option X20.

Targets may be stated for KPIs in the Incentive Schedule.

The Consultant is required to report his performance against KPIs to the Service Manager at intervals stated in the Contract Data including the forecast final measurement. If the forecast final measurement will not achieve the target stated in the Incentive Schedule, the Consultant is required to submit his proposals to the Service Manager for improving performance.

The Consultant is paid the amount stated in the Incentive Schedule if the target for a KPI is improved upon or achieved. Note that there is no payment due from the Consultant if he fails to achieve a stated target.

The Client may add a new KPI and associated payment to the Incentive Schedule but may not delete or reduce a payment.

Option Y(UK)1: Project Bank Account

In 2008, the Office of Government Commerce (OGC) published a guide to fair payment practices, following which the NEC Panel prepared a document in June 2008 to allow users to implement these fair payment practices into NEC contracts.

This is now Option Y(UK)1, which provides for a Project Bank Account to be set up which receives payments from the Client which is, in turn, used to make payments to the Consultant and to Named Suppliers. The Project Bank Account is established within three weeks of the Contract Date.

There is also a Trust Deed between the Client, the Consultant and Named Suppliers containing the necessary provisions for administering the Project Bank Account.

The Consultant should include in any subcontracts for Named Suppliers to become party to the Project Bank Account through a Trust Deed via a Joining Deed.

The Consultant notifies the Named Suppliers of the details of the Project Bank Account and the arrangements for payment of amounts due under their contracts.

The Named Suppliers will be named within the Consultant's tender, but also additional Named Suppliers may be included subject to the Client's acceptance by means of a Joining Deed, which is executed by the Client, the Consultant and the new Named Supplier. The new Named Supplier then becomes a party to the Trust Deed.

As the Project Bank Account is maintained by the Consultant, he pays any bank Charges and is also entitled to any interest earned on the account. The Consultant is also required at tender stage to put forward his proposals for a suitable bank or other entity which can offer the arrangements required under the contract.

The process every month is that the Consultant submits an application for payment including details of amounts due to Named Suppliers in accordance with their contracts.

The Client makes payment to the Project Bank Account, the Consultant makes payment to the Project Bank Account of any amounts which the Client has notified the Consultant intends to withhold from the certified amount and which is required to make payment to Named Suppliers.

The Consultant then prepares the Authorisation, setting out the sums due to Named Suppliers. After signing the Authorisation, the Consultant submits it to the Client for signature and submission to the project bank.

The Consultant and Named Suppliers then receive payment from the Project Bank Account of the sums set out in the Authorisation after the Project Bank Account receives payment.

In the event of termination, no further payments are made into the Project Bank Account.

Option Y(UK)2: The Housing Grants, Construction and Regeneration Act 1996

This option is only applicable to UK contracts where the Housing Grants, Construction and Regeneration Act 1996 applies and deals only with the payment aspects of the Act, Adjudication under the Act being covered by Option W2.

Whilst this book avoids discussing NEC4 specifically in connection with UK construction law, it is worth mentioning that the Local Democracy, Economic Development and Construction Act 2009 has a direct bearing on Option Y (UK)2.

The Local Democracy, Economic Development and Construction Act 2009, specifically Part 8, amends Part II of the Housing Grants, Construction and Regeneration Act 1996 and came into force in England and Wales on 1 October 2011 and in Scotland on 1 November 2011.

It is important to note that as the Local Democracy, Economic Development and Construction Act 2009 *amends* the Housing Grants, Construction and Regeneration Act 1996, you have to take into account *both* Acts to fully understand how it applies to your contract.

The Act applies to all construction contracts including:

- construction, alteration, repair, maintenance, etc.,
- all normal building and civil engineering work, including elements such as temporary works, scaffolding, site clearance, painting and decorating,
- Consultants' agreements concerning construction operations,
- labour-only contracts,
- contracts of any value it *excludes*,
- extraction of oil, gas or minerals,
- supply and fix of plant in process industries, e.g. nuclear processing, power generation, water or effluent treatment,
- contracts with residential occupiers,
- header agreements in connection with PFI contracts and
- finance agreements.

The HGCR Act 1996 only applied to contracts in writing, but the LDEDC Act 2009 also applies to oral contracts.

Briefly, the differences between the Housing Grants, Construction and Regeneration Act 1996 and the Local Democracy, Economic Development and Construction Act 2009 include:

- A notice is to be issued by the payer to the payee within five days of the payment due date or by the payee within not later than five days after the payment due date, the amount stated is the notified sum. The absence of a notice means that the payee's application for payment serves as the notice of payment due, and the payer will be obliged to pay that amount.
- A notice can be issued a prescribed period before the final date for payment which reduces the notified sum.
- The Consultant has the right to suspend all or part of the service and can claim for reasonable costs in respect of costs and expenses incurred as a result of this suspension as a compensation event.
- Terms in contracts such as "the fees and expenses of the Adjudicator as well as the reasonable expenses of the other party shall be the responsibility of the party making the reference to the Adjudicator" will be prohibited. Much has been written about the effectiveness of such clauses, and whether they comply with the previous Act, so the new Act should provide clarity for the future.
- The Adjudicator is permitted to correct his decision so as to remove a clerical or typographical error arising by accident or omission. Previously, he could not make this correction.

Under Option Y(UK)2, the date on which a payment becomes due is seven days after the assessment date, the final date for payment being:

- if the Service Manager makes an assessment after the defects date or the fate, the last defect is corrected, five weeks after whichever is the later;
- if the Service Manager does not make an assessment after the defects date or the fate, the last defect is corrected, one week after the Consultant issues its assessment;
- if the Service Manager has issued a termination certificate, fourteen weeks after the issue of that certificate.

The final date for payment is the later of

- fourteen days after the date on which the payment becomes due or a different period of stated in the Contract Data and
- seven days after receipt by the Party making payment of an invoice.

If either Party intends to withhold payment of an amount due, he must notify the other not later than seven days before the final date for payment by stating the amount proposed to be withheld and the reason for withholding payment.

If the Consultant exercises his right under the Housing Grants, Construction and Regeneration Act 1996 to suspend performance, it is a compensation event.

NB The publishers have issued a brief amendment to the PSC to align it with the Local Democracy, Economic Development and Construction Act 2009.

Option Y(UK)3: Contracts (Rights of Third Parties) Act 1999

The principle of privity of contract is that a person who is not a party to a contract cannot enforce the terms of that contract.

The Contracts (Rights of Third Parties) Act 1999 effectively abolishes privity of contract as an absolute principle, enabling a person who is not a party to a contract, a "third party", to enforce a term of the contract if the contract expressly provides that he may. The third party must be expressly identified in the contract either by name or as a member of a class or by a particular description, but does not need to be in existence at the date of the contract.

Under the Act, there is available to the third party any remedy that would have been available to him in an action for breach of contract as if he had been a party to the contract.

The parties to the contract are prohibited from making any agreement to vary the contract in such a way as to adversely affect the third party's entitlement to enforce the contract without the third-party's consent.

While the Act is not specially aimed at the construction industry, it is an industry which spends a lot of time and money in constructing collateral warranties enabling a multitude of parties to have certain rights over specified periods of time, so it is now possible to draft building contract documents to confer benefits upon the parties who would have had the benefit of collateral warranties.

The construction industry has generally not been in favour of this Act, preferring to use collateral warranties, but it must be said that the Act has several advantages in terms of time and cost in that only one document needs to be drafted, though it is fair to say that many parties prefer a warranty as specific evidence of their rights. Banks and funders in particular like the comfort of a good old-fashioned document.

Option Z: additional conditions of contract

Option Z allows conditions to be added to or omitted from the core clauses.

All changes to the core clauses should be included as Z clauses rather than amending the core clauses themselves; so, in effect, the clause remains in the contract, but is amended within the Z clause.

It is critical that when drafting a Z clause, it must be clearly stated what happens to the original core clause, for example, it is deleted. If the original core clause is not deleted, then it is likely that an inconsistency will arise which will be interpreted against the party who wrote or amended the clause, i.e. the Client.

These conditions may modify or add to the core clause to suit any risk allocation or other special requirements of the particular contract. However, changes should be kept to a minimum, consistent with the objective of using industry-standard, impartially written contracts.

It must be remembered that if a Client amends a contract to allocate a risk to the Contractor which may have been intended to be held by the Client, the Contractor has a right to price it in terms of time and money; therefore, the practice of Clients amending contracts to pass risk to Contractors without considering who is best able to price, control and manage those risks can in many cases prove to be unwise and uneconomical.

It is important to spend time considering whether a Z clause is appropriate in each case; then, when that decision has been made, that the clause is drafted correctly and aligned to the drafting principles of the original contract, in the case of the TSC, use ordinary language, present tense, short sentences and bullet pointing.

It is not unusual to see Z clauses in a Professional Service Contract written in a legalistic language, in the future tense and without punctuation apart from full stops!

The Z clause provision allows additional conditions to be added to further tailor the contract strategy including addition or deletion of core clauses. It may include adding or deleting compensation events. The flexibility inherent in the Professional Service Contract is designed so that amendments to the contract are kept to a minimum.

However, those doing so should:

- be aware of how it might impact on the procedures within the contract
- maintain the principles of the PSC
- not fill the contract with minor issues which are unlikely to arise
- not undermine the stimulus to good management
- not undermine the drafting philosophy of the PSC by attempting to prescribe an outcome for every foreseeable eventuality
- seek to maintain the principle of managing the problems

Should it be necessary to amend standard clauses, all the changes must be shown under Option Z. This ensures that all the changes are highlighted.

A correctly drafted example

Payment

The Consultant's share
Delete Clause 54.3
Insert:

The *Service Manager* makes a preliminary assessment of the *Consultant's* share at any assessment date. If the forecast Price for Work Done to Date is less than the final total of the Prices, this share is included in the amount due following completion of the whole of the service. If the forecast Price for Work Done to Date is more than the forecast total of the Prices, this share is retained from the amount due.

Allocating risk using the Main and Secondary Options

Before deciding on a contract strategy, the project objectives, constraints and risks should be identified. Analysis enables those factors to be appropriately allocated.

The fundamental principles of risk allocation are:

- risks should be allocated to the Party best able to control them prior to their occurrence,
- risks should not be allocated to a Party who is unable to sustain their consequences,
- risks which are outside the Consultant's control should usually be allocated to the Client,
- risk allocation should encourage good management by the party who carries the risk,
- motivation regarding risk is provided by the financial consequence,
- where risk allocation is split, the split should not only reflect each party's ability to influence the likelihood of occurrence and effect and
- the contract should not be cluttered up with risks of small likelihood and impact.

It is now accepted that the interests of the Client are best served by limiting the risks carried by the Consultant to those he can control better than the Client. The PSC, through its Main and Secondary Options, allows the Client to implement a policy of risk allocation which maximises the benefit and minimises the effects of time and cost.

The PSC uses risk allocation to encourage good management of the project with the risks allocated to the Party most able to respond to them.

There are two opportunities to manage risk allocation: first, by the choice of the contract Main Option and, second, by making an appropriate allocation of risk by the choice of the Secondary Options.

Contract strategy

As stated above, construction procurement today indicates a movement away from the traditional use of bills of quantities for tendering and paying Contractors, with a movement towards payment mechanisms such as milestone payments, activity schedules and payment schedules with payment based on progress achieved against targets rather than merely measuring and paying for the quantity of work done.

This trend is also true with Consultants who are being employed less on a time based or ad valorem basis and more on achievement of progress through milestones, activity schedules, etc.

In both cases, there is also the increasing growth of the use of target cost contracts which have been encouraged by the increasing use of partnering and alliancing arrangements to incentivise the Parties to create savings for joint benefits.

There is also an increasing tendency for greater Consultant involvement with Contractors through the increasing use of design and build/design and manage contracts, performance specifications and PFI/PPP design, build, finance and operate projects.

Contract strategy decisions

The Client usually considers what might be the most appropriate contract strategy during the feasibility stage and before any design or construction has been undertaken.

Factors to take into account in deciding which of the Main Options to use from within the PSC include the following:

- how clearly identified are the Client's requirements,
- what is the likelihood of change to those identified requirements,
- how important to the Client is certainty of price,
- what views prevail on the allocation of risk,
- which Party is to be responsible for design,
- how important is early commencement and/or rapid completion and
- does the Client need flexibility in the contractual arrangements.

The result should be a statement of the chosen contract strategy and the range of Main and Secondary Options to be used to achieve that.

The advantage of using the PSC is that whatever variations in strategy between different contracts within a project are adopted, the majority of the procedures, based upon the Core Clauses, will be common to all contracts.

Standard of care of consultants

Professional appointments will set the standards required from a Consultant.

A basic standard of care is to state that the Consultant will use "reasonable skill and care" in performing its services.

This standard however is not considered to be the "highest possible" degree of skill and care, as the Consultant will be judged and measured against an average consultant in the same profession.

The contract might therefore strengthen the standard of care by, for example, referring to a Consultant having the appropriate qualifications for a particular industry and/or being an experienced Consultant on projects of a similar size to the project in question. This ensures that the Consultant is tied to measurable standards.

Reasonable skill and care can have a higher threshold in cases where a Consultant has held himself out as having particular skills and abilities above what would be considered as "normal".

Words in a contract that would raise the standard of care above normal are indicating that the Consultant is an "expert", "competent" or a "specialist" in a particular field.

A stronger standard is "fitness for purpose". This means that when the project is completed, it will be fit for its intended use.

It means that the Consultant is effectively guaranteeing that, for instance, its design will be fit for purpose. Consultants will often attempt to resist a fitness for purpose obligation, as it is possible that it will lead to them having difficulties covering themselves under their PI insurance.

To illustrate the difference between the two standards, if an Architect designs a hospital which cannot be used as a hospital once construction is complete:

- if the design standard is "reasonable skill and care", as long as the Architect can prove that it used reasonable skill and care in its design, it will not be liable regardless of the hospital not being fit for purpose.
- if the design standard is "fitness for purpose": even if the Architect can prove that it used reasonable skill and care in its design, it will be liable for the hospital not being fit for purpose.

Design and build contracts

The Consultant may in some cases be employed by the Contractor under a Design and Build arrangement.

There are a number of reasons for allocating design to the Contractor:

- the design and construction periods can overlap, leading to faster delivery of the project;
- the Contractor can utilise his particular experience and preferred methods of construction to minimise costs and price;
- the temporary works/permanent works interface and influence of design is rationalised, and the permanent works should be more "buildable";
- the traditional design/construction interface and its recognised problems are eliminated;
- the management of the design risk by the Contractor can result in greater certainty of the time, cost and performance and project objectives being met.

Novation of designers

Whilst the principle of design and build agreements is that the Client prepares his requirements and sends them to the tendering Contractors, and they prepare their proposals to match the Client's requirements; in reality, in nearly half of design and build contracts, the Client has already appointed a design team which prepares feasibility proposals and initial design proposals before tenders are invited. Outline planning permission and sometimes detailed permission may have also been obtained for the scheme before the Contractor is appointed.

Each Contractor then tenders on the basis that the Client's design team will be novated or transferred to the successful tendering Contractor who will then be responsible for appointing the team and completing the design under a new agreement.

This process is often referred to as "novation", which means "replace" or "substitute" and is a mechanism where one party transfers all its obligations and benefits under a contract to a third party. The third party effectively replaces the original party as a party to that contract, so the Contractor is in the same position as if he had been the Client from the commencement of the original contract. Many prefer to use the term "consultant switch" where the design consultant "switches" to work for the Contractor under different terms as a more accurate definition.

This approach allows the Client and his advisers time to develop their thoughts and requirements, consider planning consent issues, then when the design is fairly well advanced, the Designers can then be passed to the successful design and build Contractor.

The NEC4 contracts do not include pro forma novation agreements, but, if used, it is critical that the wording of these agreements be carefully considered as there are many badly drafted agreements in existence. Novation can be by a signed agreement or by deed. As with most contracts, there should be consideration, which is usually assumed to be the discharge of the original contract and the original parties' contractual obligations to each other. If the consideration is unclear, or where there is none, the novation agreement should be executed as a deed.

In many cases, the agreement states briefly (and very badly) that from the date of the execution of the novation agreement, the Contractor will take the place of the Client as if he had employed the Designers from the beginning, the document stating that in place of the word 'Client', one should read 'Contractor'; however, the issues of design liability, inspection, guarantees and warranties may not always apply on a back-to-back basis, so one must take care to draft the agreement in sufficient detail and refer to the correct parties.

In practice, the best way is to have an agreement drafted between the Client and the Designer covering the pre-novation period, and a totally separate agreement drafted between the Contractor and the Designer for the post-novation period.

When a contract is novated, the other (original) contracting party must be left in the same position as he was in prior to the novation being made. Essentially, a novation requires the agreement of all three parties.

As this book is intended as a guide for NEC4 practitioners worldwide, it is not intended to examine specific legal cases within the UK, as they may not apply on an international basis, but there is certainly case law available in terms of design errors in the pre-novation phase being carried through as the liability of the Contractor in the post-novation stage, and also the accuracy and validity of Site investigations and the Client's and the Contractor's obligations in terms of checking and taking ownership of such information.

The roles of the parties in the NEC4 Professional Service Contract

The parties referred to in the NEC4 PSC are:

- the Client,
- the Consultant,
- the Service Manager and
- the Adjudicator.

The Client

Whilst the NEC3 Professional Services Contract referred to "the Employer", the NEC4 Professional Service Contract refers to "the Client", and this terminology has been adopted throughout the NEC4 Contracts.

Although the contract is between the Client and the Consultant, the Client is rarely mentioned in the contract, the Service Manager dealing with the various actions under the Contract.

The Client's obligations are:

- to act as stated in the contract and in a spirit of mutual trust and co-operation;
- to allow the Consultant access to and use of each part of the Site (if required) for the Consultant to carry out the service included in the contract. This access must be allowed on or before the later of the access date (the date in the Contract Data), and the date for access is shown on the Accepted Programme;
- to allow the Consultant access to and use of a part of the Site (if required) which he has taken over if they are needed for correcting a Defect;
- to make payment to the Consultant in accordance with the Service Manager's certificates and
- to take over the service at Completion.

The Consultant

The Consultant and the Client are the Parties to the Contract, the Consultant's responsibility being to Provide the Service in accordance with the Contract, the Scope describing the services and any constraints on how he has to provide it.

There are also obligations in respect of complying with instructions, notifying early warnings, submitting programmes for acceptance and notifying and submitting quotations for compensation events which are highlighted in the later chapters.

The Service Manager

Under the NEC3 Professional Services Contract, the interactions were between the Employer and the Contractor, but under Secondary Option X10, if the Employer wished, he could appoint an Employer's Agent to act for him.

Under the NEC4 Professional Service Contract, the Service Manager is appointed by the Client and manages the contract on his behalf.

The Service Manager is *required* to be appointed from the start; it is *not* an option.

The Service Manager is named in Contract Data Part 1 and may be appointed by the Client's own staff or may be an external consultant. The Service Manager can be a named individual or a company that then nominates an individual.

It is vital that the Client appoints a Service Manager who has the necessary knowledge, skills and experience to carry out the role, which includes issuing early warnings, acceptance/non-acceptance of plans, certifying payments and dealing with compensation events.

Disciplines that have carried out the Service Manager role to date include Engineers, Architects, Building Surveyors and Quantity Surveyors, in addition to Service Managers themselves.

It is also vital that the Client gives the Service Manager full authority to act for him, particularly when considering the time scales imposed by the contract. If the Service Manager has to seek approvals and consents from the Client, then he must do so, and the appropriate authority must be given, in compliance with the contractual timescales. If the approval process is likely to be lengthy, it is critical that both the Client and his Service Manager set up an accelerated process to comply with the contract or that the time scales in the contract are amended. The former is the vastly preferred method in order that NEC4 will work to its full effect.

Although the Service Manager manages the contract at the post-contract stage, he is usually appointed pre-contract to deal with matters such as feasibility issues, advising on procurement, cost planning, tendering and programme matters.

Note that, unlike many other contracts, the PSC does not name a Quantity Surveyor, so this would need to be considered in dealing with financial matters pre and post-contract. Either the Service Manager may carry out the role of the Quantity Surveyor himself or delegate to another.

There should only be one Service Manager for the project, though the Service Manager can delegate responsibilities to others after notifying the Contractor and may subsequently cancel that delegation (Clause 14.2).

As with any other notification under the contract, if the Service Manager wishes to delegate, the notice to the Contractor must be in a form which can be read, copied and recorded, i.e. by letter or email, not by verbal communication such as a telephone call.

The delegation may be due to the Service Manager being absent for a period due to holidays or illness or because he wishes to appoint someone to assist him in his duties.

Although the contract is not specific, the notice should identify who the Service Manager is delegating to, what their authority is and how long the delegation will last, so the Contractor is in no doubt as to who has authority under the contract.

It is important to recognise that in delegating, the Service Manager is sharing an authority with the delegate who will then represent him, and he is not passing his responsibilities on to them. The authority is shared, but ultimately, the responsibility will remain with the Service Manager as the principal.

It is also important to note that one can only delegate outwards from the principal, i.e. no-one can assume authority possibly because, within an organisation, they are senior to the Service Manager.

If the Client wishes to replace the Service Manager, he must first notify the Contractor of the name of the replacement before doing so.

The Adjudicator

The Adjudicator can be chosen by one of the three methods:

- By the Client specifically naming the individual in Contract Data Part 1. The Consultant therefore knows at the time of tender who the Client has named and can therefore confirm that he does not agree with the Client's choice. This has advantages in that the parties always know who the Adjudicator will be if they have a dispute. The disadvantages are first that NEC practitioners have commented in the past that naming the individual within the Contract Data assumes that there will be a dispute, and often the Adjudicator will require payment for having his name in the contract and being available should the parties have that dispute.
- By the Client naming an Adjudicator nominating body in Contract Data Part 1. This would normally be a professional institution. His tends to be the favoured method.
- By the parties agreeing who will be the Adjudicator at the time of the dispute. The disadvantage of this method is that once the parties are in dispute, they probably would not readily agree with each other on anything.

The Adjudicator becomes involved only when either contracting party refers a dispute to him. As an impartial person, he is required to give a decision on the dispute within stated time limits. If either party does not accept his decision, they may refer the dispute to the tribunal (litigation or arbitration). The Adjudicator's Contract requires that payment of the Adjudicator's fees is shared by the Parties unless otherwise stated.

Mutual trust and co-operation

The first clause of all the NEC contracts (Clause 10.1) has always required the Parties and their agents, e.g. Project Manager, Supervisor and now Service Manager, to act in a spirit of mutual trust and co-operation.

This has been changed within the NEC4 contracts in that the former Clause 10.1 has now been split into two separate clauses: Clauses 10.1 and 10.2:

Under the NEC4 Professional Service Contract:

- Clause 10.1: The Parties and the Service Manager shall act as stated in the contract.
- Clause 10.2: The Parties and the Service Manager act in a spirit of mutual trust and co-operation.

If we examine the clauses, first, with Clause 10.1, is it necessary to state that the Parties and the Service Manager are required to act as stated in the contract? What difference does it make if that clause was absent or deleted? Would they not have to act as stated in the contract?

Second, within Clause 10.2, what does it mean that the Parties and the Service Manager are required to act in "a spirit of mutual trust and co-operation"?

This second requirement mirrors Sir Michael Latham in his report "Constructing the Team" when he recommended that the most effective forms of contract should include "a specific duty for all parties to deal fairly with each other and in an atmosphere of mutual co-operation".

This provision has often been viewed with some confusion, and for those who have spent many years in the construction industry, with a degree of scepticism.

Most practitioners state that their understanding of the clause is that the parties should be non-adversarial towards each other, acting in a collaborative way and working for each other rather than against each other, in reality, that is what the clause requires and is how parties should behave towards one another anyway.

However, the difficulty is that if a party does not act in a spirit of mutual trust and co-operation, what can another party do? How would you know, and what is the remedy?

The answer is that the clause is almost unenforceable as it is virtually impossible to define "mutual trust and co-operation" then to quantify the breach or the ensuing damages that flow from the breach. In addition, the Contractor is not contractually related to the Service Manager, so neither would be able to take action directly against the other for breach of contract, it would have to be via their mutual relationship with the Client.

To that end, as Clause 10.1 is a fairly redundant clause, and Clause 10.2 is unenforceable, Clients have been seen to insert a Z clause to delete the requirements; however, having raised a concern, it is recommended that it should remain in the contract, if merely viewed as a statement of good intent.

In effect, a clause requiring parties to act in a certain spirit will probably not, on its own, have any real effect. Within the PSC, it is the clauses that follow within the contract which require early warnings, clearly detailed plans and programmes which are submitted for acceptance, and a structured change management process that actually create and develop that level of mutual trust and co-operation rather than simply inserting a statement within the contract requiring the parties to do so.

Some unique NEC4 Professional Service Contract terms

There are some terms that, whilst not being used in other contracts, have different meanings within NEC4 contracts.

Others

The PSC uses the term "Others" to define parties who are "not the Client, the Service Manager, the Adjudicator, the Consultant, or any employee, Subconsultant or supplier of the Consultant". Essentially, Others are anyone outside the contract, for example, other consultants, statutory bodies, regulatory authorities, utilities companies, etc.

References to Others include the Consultant's obligation to working with Others in obtaining and providing information (Clause 22.1) and to obtain approval from Others where necessary to Provide the Service (Clause 24.1).

Communications – clause 13

The NEC4 family has strict rules regarding communications under the contract, and this is reflected within the PSC:

- "Each communication which the contract requires must be communicated in a form which can be read, copied and recorded". This may be by electronic means, for example, email or via a project intranet, so the issue and receipt are simultaneous. Normally, the parties agree a protocol for project communications.
- This requirement is particularly important in respect of instructions as the NEC4 contracts do not recognise verbal instructions and whereas under other forms of contract, a verbal instruction can be confirmed by the Consultant and if not dissented from, normally within seven or fourteen days, it becomes an instruction, and the NEC4 contracts do not have that provision. The instruction may be in the form of a letter, a pro forma instruction, an email or any other written communication.
- As the NEC4 contracts are intended for international use, the language of the contract is stated within Contract Data Part 1, and all communications should be in that language.
- The Client and the Consultant reply within the "period for reply" stated in Contract Data Part 1, unless the contract states otherwise. This period would normally be seven or fourteen days, but if the Client and the Consultant agree, the Client may extend the period for reply.
- If the Client is required to accept or not accept, he must state his reasons for non-acceptance, and this provides clarity in allowing the Consultant to know what he needs to do to get acceptance by the Client.

  Withholding acceptance for a reason stated in the contract is not a compensation event, and this is particularly in respect of:

  i Acceptance of a proposed key person (Clause 21.1);
  ii Acceptance of a proposed Subcontractor (Clause 23.2);
  iii Acceptance of a Consultant's programme (Clause 31.3).

Withholding acceptance because the Client requires more information in order to assess the Consultant's submission fully is a reason and therefore not a compensation event.

The Consultant is required to retain copies of drawings, specifications, reports and other documents for the period of retention and in the form stated in the Scope. Typically, the period of retention would be six or twelve years.

Most countries have a statutory limitation period for latent Defects which come to light at a later date, beyond which a claim cannot be made for defective work.

In the UK, for example, the time limit for actions founded on simple contract is six years, and for a specialty contract or deed, the time limit is twelve years; in both cases, this is measured from the date on which the cause of action accrued; normally, this is taken as the date of Completion.

Notifications should be communicated separately from other communications, so, for example, an early warning notice should not be included as part of a set of meeting minutes or a letter which includes other subjects.

Note

1 Latham Report (1994) "Constructing the Team": Joint Review of Procurement and Contractual Arrangements in the United Kingdom Construction Industry. Chair Sir M. Latham. HMSO, London.

1 Early warnings

Introduction

Early warnings are covered within the NEC4 Professional Service Contract by Clauses 15.1–15.4.

Early warnings are an essential and valuable risk management tool within the NEC4 contracts, the Professional Service Contract obliging the Consultant and the Service Manager to notify each other as soon as either becomes aware of any matter which could affect the service in terms of time, cost or quality.

Under Clause 15.1, the Consultant and the Service Manager give an early warning by notifying the other as soon as either becomes aware of any matter which could:

- increase the total of the Prices.
  The Prices of the service, in the form of the activity schedule, the target or the forecast Defined Cost.
- delay Completion.
  The Completion of the whole of the service.
- delay meeting a Key Date.
  Delay the completion of an intermediate "milestone date" in accordance with Clause 11.2(9).
- impair the usefulness of the service to the Client.
  This sometimes causes confusion, but it could be the Consultant notifying the Service Manager that a survey he is carrying out is no longer relevant or needs to be amended, or the Service Manager could notify the Consultant that he is perhaps not carrying out the survey properly and the deliverables are not as required.
- affect the work of the Client, a Client's contractor or another consultant.

The Consultant may also give an early warning by notifying the Service Manager of any other matter which could increase its total cost.

One could query whether a matter which could increase the Consultant's cost, but not affect the Prices, should be an early warning matter, or for that matter, whether it should be anything to do with the Service Manager, particularly if Option A has been selected, so the Consultant is not looking to recover this additional cost through the contract, but the words are "the Consultant may give an early warning" so he is not obliged to do so.

The Consultant is not required to give an early warning for which a compensation event has previously been notified, so, as an example, if the Service Manager gives an instruction which changes the Scope, it is a compensation event, for which neither the Service Manager or the Consultant is required to give early warning.

The early warning procedure obliges people to be "proactive", dealing with risks as soon as the parties become aware of them rather than "reactive" waiting to see what effect they have then trying to deal with them when it is often too late. Encouraging the early identification of problems by both parties puts the emphasis on joint solution finding rather than blame assignment and contractual entitlement.

It is one of the most important and valuable aspects of the contract, and it is perhaps surprising that, whilst few other contracts refer to an early warning process, only the NEC contracts set out in clear detail what the parties are obliged to do, with appropriate sanctions should the parties not comply (see Clauses 61.5 and 63.7).

Notifying early warnings

The contract requires (Clause 13.1) that communications that the contract requires are in a form that can be read, copied and recorded, so early warnings should not be a verbal communication such as a telephone conversation. If the first notification is a telephone conversation, or a comment in a site meeting, it should be immediately confirmed in a form that can be read, copied and recorded as an early warning to give it contractual significance (see Figures 1.1 and 1.2).

Also, Clause 13.7 requires that notifications which the contract requires must be communicated separately from other communications; therefore, early warnings must not be included within a long letter which covers a numbers of other issues which may or may not be related to the specific issue or embodied within the minutes of a progress meeting.

There are some key words within the obligation to notify:

- "*The Consultant and the Service Manager*" – No one else has the authority or obligation to give an early warning. The Service Manager is therefore notifying on behalf of himself and the Client, and any possible others who he may represent within the contract. The Consultant is notifying on behalf of himself, his Subcontractors (if appropriate), and again many possible others who he represents under the contract. Early warnings should be notified by the key persons named in Contract Data Part 2.

Service Managers are often criticized for seeing early warnings as something the Consultant has to do, and in fact most early warnings are actually issued by the Consultant. However, the Consultant and the Service Manager are obliged to give early warnings each to the other, so it is critical that Service Managers play their part in the process.

As an example, if the Service Manager becomes aware that he will be late in delivering some information to the Consultant, he should issue the early warning as soon as becomes aware that the information will not be delivered to the Consultant, not wait and subsequently blame the Consultant for not giving an early warning stating that he has not received the information!

- "As soon as" – means immediately. Clauses 61.5 and 63.7 deal with the situation where the Consultant did not give an early warning. Whilst the party who gives the early warning must do so as soon as he becomes aware of the potential risk, the other party should respond as soon as possible and in all cases within the period for reply stated in Contract Data Part 1.

Contract:	EARLY WARNING NOTICE
Contract No:	EWN No.......................................

Section A: Enquiry

To: Service Manager/Consultant

Description

This matter could:

- ☐ Increase the total of the Prices
- ☐ Delay Completion
- ☐ Delay meeting a Key Date
- ☐ Impair the usefulness of the service to the Client
- ☐ Affect the work of the Client, an Client's Contractor or another consultant

Early warning meeting called? Yes/No Date: ....

Signed:................................(Service Manager/Consultant) Date: ....

Action by: Date required:

Section B: Reply

To: Service Manager/Consultant

Signed:................................(Service Manager/Consultant) Date: ....

Copied to:

Service Manager ☐ Consultant ☐ Other .... ☐ File ☐

Figure 1.1 Sample early warning notice.

- "Could" – not must, will or shall. Clearly, there is an obligation to notify even if it is only felt something *may* affect the contract, but there is no clear evidence that it will.

Contract: .... **Contract No:**				**RECORD OF EARLY WARNINGS**				
Ref.	Date notified	Origin SM or Cons	Description of event	Clauses	Ref.	Action by SM or Cons	Date Closed Out	Notes
1								
2								
3								
4								
5								
6								
7								
8								
9								
10								
11								
12								
13								

Service Manager ☐ Consultant ☐ Other .... ☐ File ☐

Figure 1.2 Sample record of early warning notices.

The Service Manager enters early warning matters in the Early Warning Register.

It must be emphasised that early warnings are not the first step towards a compensation event or even a dispute, as is often believed.

Early warnings feature in a completely separate section of the contract, and in fact, the early warning provision is intended to prevent a compensation event occurring or at least to lessen its effect. It can also be used to notify a problem which is totally the risk of the notifier. It is also worth mentioning that early warnings are a notice of a future risk, not a past one.

The parties are not required, nor is it of any value, to notify a risk that has already happened.

Remedy for failure to give early warning

Under Clauses 61.5 and 63.7, if the Service Manager decides that the Consultant did not give an early warning of the event which an experienced consultant could have given, he notifies this to the Consultant when he instructs him to submit quotations. If the Service Manager has done so, the event is assessed as if the Consultant had given early warning.

Example

On 1 February, the Consultant becomes aware of an issue which could increase the total of the Prices and delay Completion. He should immediately give an early warning to the Service Manager; however, he fails to do so until 22 February, three weeks later.

When the Consultant gives the early warning, the Service Manager notifies a compensation event and instructs the Consultant to submit a quotation, at the same time notifying the Consultant that he did not give an early warning of the matter that an experienced Consultant could have given. If the Service Manager has given such notification, then the compensation event is assessed as if the Consultant had given early warning, so the Consultant should price his quotation based on the event as on 1 February rather than on 22 February.

Defining risk

Risks can be classified in a Professional Service Contract as:

- i Risks which are owned by the Client and do not affect the Consultant.
- ii Risks which are owned by the Client and do affect the Consultant in terms of Price, Key Dates and/or Completion Date and are therefore included and managed as compensation events.
- iii Risks which are owned by the Consultant and are therefore deemed to have been included within his Price and programme. These are risks which may affect the Consultant, but are not listed as compensation events.

Dependent on the main option chosen, a risk held by the Consultant may still be paid by the Client, even though it will not be a compensation event.

General principles of risk management

There are a number of definitions of risk, an example being:

> Any uncertain event or set of circumstances that, should it or they occur, would have an effect on achieving the objectives.
>
> (Association for Project Management)

The objectives when engaging a Consultant under a professional services contract are that they will provide a service, to the required quality and standards, they will work safely and they will complete within the required time scale and within the required budget.

Risk is always inherent in any activity, whether it be consulting or not, and no matter how simple or complex it may be. However, the degree of risk will vary depending upon the circumstances involved, and the amount of risk that one is willing to assume is normally directly proportional to the amount of reward that is anticipated. The greater the anticipated reward, the greater the risk one is normally willing to assume.

Risk is an integral part of the consulting process and an important factor in the determination of cost and duration.

The parties to a professional services contract assume that the contract will not only provide them with adequate protection but also provide a fair and workable framework within which sometimes, a vastly complex service can be provided to the mutual satisfaction of all concerned.

The contract itself should be clear, concise, complete, legally correct and equitable. It should address the risks directly and attempt to eliminate as much of the ambiguity as possible.

Characteristics causing professional service contracts to present risks

All consultancy agreements can be regarded as having risk as an underlying factor, though there are a number of differences between a professional services contract and a manufacturing process, in that, every construction project:

- i is different, the only similarities being the people who manage them.
- ii is not produced in a factory environment, but on a site which is subject to fairly predictable seasonal variations, but unpredictable weather exceptions.
- iii brings a temporary multi-skilled and experienced team together which is formed and brought together for the purpose of constructing the project and possibly never working together again.

As a result of the above, one can easily draw the conclusion that:

- i it is impossible to provide a service, whether it be, for example, a design or a survey, which will not require subsequent modifications, either because of Client changes or because the design is found in practice not to work.
- ii it is impossible for a Consultant to produce a tender which will correspond exactly to the cost plus overheads and profit which will actually be incurred when providing that service.

iii it is impossible for a Consultant to produce a programme which can in all respects apply without revision and updating.
iv when people are faced with such uncertainty, they may, in an effort to defend their own beliefs and organisations, engage in conflict situations.

As a consequence, it is clear to see that both Client and Consultant organisations are faced with risks that occur as a result of the nature of the services.

From the moment the decision to begin a service is taken until the service is completed, the Client and Consultant take on-board risks.

In addition to inherent risks within the provision of the services, there are also risks that will arise from the agreement between the parties. These contractual risks are primarily expressed within the contract as intended between the parties, but they can also arise through inexperienced drafting which can lead to lack of contract clarity, ambiguities and inconsistencies.

The main principles of apportioning risk in professional services contracts are:

- the Client will always pay for risk either by retaining the risk within the contract and paying for it should it occur or the Consultant prices it within his tender,
- they should be apportioned to the party best able to manage and/or control them and to sustain the consequences if risks materialise,
- risks outside the Consultant's control and/or foreseeability should remain with the Client, and
- the party carrying the risk should be motivated to manage the risk, possibly through a risk/reward process.

The major types of risks that can arise in professional services contracts are:

Technical
- incomplete or incorrect information or design possibly produced by others
- inadequate site investigations, possibly carried out by others

Logistical
- availability of resources
- availability of transportation in isolated locations

Resources
- insufficient resources and expertise
- industrial relations problems

Timing
- inability to give access or possession when required

Financial/Contractual
- inflationary costs
- delays in payment
- non-availability of funds/cash flow problems
- ambiguities and/or inconsistencies in contract documents
- work carried out by directly employed parties
- losses due to default of Consultants, Contractors and/or other parties

Safety/Reliability
- delays due to having to satisfy requirements

Quality Assurance/Standards
- quality issues

Choice of Site/Physical
- ground conditions
- ground water conditions
- soil strata type and variability
- contamination
- exceptional weather conditions
- damage to neighbouring buildings

Environmental and Planning
- rights of way
- public enquiries
- planning requirements
- noise abatement
- availability of services

Political
- changes in Government
- new legislation
- customs and import restrictions

Design
- suitability and adaptability of the design

Unfamiliarity
- "pioneer" design
- experience of design team

Errors in tender documents
Industrial Action
Changes
- effect on duration and Price of late changes
- effect of number of changes

Construction Delays
- delay by Consultants, Contractors and/or statutory authorities
- delay in giving instructions to the Consultant

Force majeure
- fundamental risk, which cannot be foreseen, quantified, priced or programmed

The risk management process

Risk management is a structured approach to identifying, assessing and managing risk. It can be defined as:

> The systematic application of management policies, procedures and practices to the tasks of identifying, analysing, assessing, treating and monitoring risks.

The risk management process can help a party analyse the degree of uncertainty in achieving the contract/project objectives, understand the consequences of particular events occurring and develop management actions to control the event or minimise its consequences.

There are six primary activities in the process:

1 Risk Identification

Risks should be identified as early as possible and continuously monitored to highlight new risks and the passing of old risks. Risk identification should initially be carried out by analysing all available documentation and knowledge as a brainstorming session involving a number of individuals, ideally chaired by an independent person external to the project.

2 Risk Analysis

Once the risks have been identified, they should be classified in terms of:

- likelihood of occurring
- impact on the project objectives should they occur.

Each risk can then be given a ranking, possibly the likelihood in ascending order:

- Extremely unlikely
- Unlikely
- Moderate
- Likely
- Almost certain

Impact can also be given a ranking, possibly the impact in ascending order:

- Insignificant
- Moderate
- Major
- Severe
- Catastrophic

3 Risk Planning

This is the process of developing responses to address, manage or control risks.

The responses will consider:

- *Avoiding the risk*: the Client abandoning the project, the Consultant deciding not to tender?
- *Reducing the probability*: doing something a different way or educating the people that may be subject to the risk, so the risk is less likely to occur?
- *Reducing the impact*: including resources, funds and/or float in programmes to cushion the effect of the risk?
- *Transferring the risk*: passing the risk to another (willing) party or an insurance company who is best able to manage or bear the risk?
- *Accepting the risk*: with the *full* knowledge of what the risks and potential outcomes are.
- *Ignoring the risk*: but only with the *full* knowledge of what the risks and potential outcomes are.

4 Risk Tracking

Risks should be tracked throughout the project to monitor the status of each risk as the project progresses.

5 Risk Controlling

An activity that utilises the status and tracking information about a risk or risk mitigation effort, a risk may be closed or watched, and a contingency plan may be involved.

6 Risk Communicating
An action to communicate and document the risk at all times. This will normally be done in the form of a risk register.

When considering risk, it is vital to consider:

- Which party can best foresee the risk
- Which party can best bear the risk
- Which party can best control the risk
- Which party most benefits or suffers if the risk materialises

The use of an Early Warning Register under the NEC4 Professional Service Contract

Under the NEC3 Professional Services Contract, the "Early Warning Register" was called the "Risk Register", but essentially, it was not a Risk Register in the normal sense, so the name change is welcomed.

The Contract Data does contain matching sections in Parts 1 *and* 2 for the Parties to add matters that will be included in the Early Warning Register.

That is to say that both Parties have the ability to list risks that they wish to form part of the risk management processes in the contract. These, as we will see, *do not* change the risk allocation between the Parties. These entries are to assist the risk management processes of the Parties by listing those risks which collaborative behaviour will help with.

It is critical that the parties fully understand that the purpose of an Early Warning Register is to list all the identified risks and the results of their analysis and evaluation. It can then be used to track, review and monitor risks as they arise to enable the successful completion of the project. The Early Warning Register does not allocate risk that is done by the contract.

The Early Warning Register is a simple document, but a vital one in the process of risk management. Its role and contents are frequently misunderstood by Parties particularly where people confuse its use with risk registers elsewhere, such as those that are components of company management systems. Its purpose is to assist the Parties and the Service Manager with managing risk in the project delivery and not to allocate responsibility or blame.

It is helpful to start with its definition from clause 11.2(7);

> The Early Warning Register is a register of matters which are
>
> - listed in the Contract Data for inclusion and
> - notified by the Service Manager or the Consultant as early warning matters.

It includes a description of the matter and the way in which the effects of the matter are to be avoided or reduced.

So, read the definition carefully. The Early Warning Register contains information about these risks:

- Those listed in the Contract Data
  - by the Client in Contract Data Part 1
  - by the Consultant in Contract Data Part 2

- Those notified as an early warning matter
  - by the Service Manager
  - by the Consultant

The information contained in the Early Warning Register is

- a description of the matter
- the way in which the effects of the matter are to be avoided or reduced

The Early Warning Register need not contain anything else. It is a document produced after the Contract Date, and therefore it does not form part of the contract.

For practical reasons though, it should not contain any statements that purport to allocate risk to one party or the other that will just prove confusing in the long run.

The contract does not prescribe the format or layout of the Early Warning Register other than to list the risks in the contract and those that come to light at a later date and are notified as early warnings following which, if there is an early warning meeting, the Service Manager revises the Early Warning Register to record the outcome of the meeting.

Contract Data Part 2 allows the Consultant to identify matters which will be included in the Early Warning Register. The Early Warning Register does not allocate or change the risks in the contract, and it records them and assists the parties in managing them. In that sense, it is a valuable addition to the workings of the contract.

In effect, the Consultant could seek to qualify its tender by including additional matters within Contract Data Part 2 which it sees or would like to transfer as Client's liabilities.

That is not the intent of the Professional Service Contract as the only place to amend the risk profile is in Contract Data Part 1.

However, the wording of the final bullet of Clause 80.1 leaves open the possibility of the Consultant inserting such an entry into Contract Data Part 2. Clients should check for such entries during tender appraisal.

As stated previously, the Early Warning Register does not allocate or change the risks in the contract, and it records them and assists the parties in managing them. In that sense, it is a valuable addition to the contract which was absent from previous editions but introduced as a Risk Register in NEC3.

Risk management is essential to the success of any project and normally follows five steps:

Step 1: Identify the risks
Step 2: Decide who could be harmed and how
Step 3: Evaluate the risks and decide on precautions
Step 4: Record the findings and implement them
Step 5: Review the assessment and update if necessary

The contract does not prescribe the format or layout of the Early Warning Register or its intended purpose, other than to list the risks in the contract and those that come to light at a later date and are notified as early warnings following which, if there is an early warning meeting, the Service Manager revises the Early Warning Register to record the outcome of the meeting.

EXTRACT FROM EARLY WARNING REGISTER

Contract:........................

Contract No:....................

Client.........................
Service Manager......................
Consultant.........................

Early Warning No. 12 - Design of foundations to new office block

Description of Risk	Implications	Likelihood of Occurrence (1 - 5) (Least - Most)	Potential Impact (1 - 5) (Low - High)	Risk Score	Risk Owner	Mitigation Strategy By whom? By when?	Allowance in the total of the Prices	Programme Allowance	Client cost allowance	Risk Status	Last Updated
Possibility of unforeseen physical conditions affecting foundation design	(i) Possible part redesign by Consultant (ii) Delay and additional costs in removal by Contractor	3	4	12	Client Ground investigations carried out by Client appointed geotechnical consultant pre contract may not be sufficient	Suggest additional investigations be instructed before design progresses and Contractor appointed	No allowance for unforeseen physical conditions	No allowance for unforeseen physical conditions	£75,000	Current	5/1/2020

Figure 1.3 Extract from early warning register.

In order to compile the Early Warning Register, the risks must first be listed, and then they are quantified in terms of their likelihood of occurrence and their potential impact upon the project.

Figure 1.3 shows an extract from a typical Early Warning Register which would comply with the contract in that it includes the basic requirement for a description of the risk and a description of the actions which are to be taken to avoid or reduce the risk.

There is no stated list of components of an Early Warning Register, but column headings should typically be titled:

i Description of risk

A clear description of the nature of the risk, if necessary referring to other documents such as site investigations, etc.

ii Implications

What would happen if the risk were to occur?

iii Likelihood of occurrence

This provides an assessment of how likely the risk is to occur. The example shows a forecast on a 1 (least likely) to 5 (most likely) basis, though it may be assessed as percentages, colour coding or simply "Low" (less than 30% likelihood), "Medium" (31–70% likelihood) and "High" (more than 70% likelihood).

iv Potential impact

This assesses the impact that the occurrence of this risk would have on the project in terms of time and/or cost. The example shows the assessment on a 1 (low) to 5 (high) basis.

v Risk score

Risk = Likelihood of something occurring × Impact should it occur.

By evaluating risk on that basis, each can be evaluated and categorised into an order of importance. This formula can only be applied to economic loss, and different standards would need to be adopted when considering risk of death or serious bodily injury.

vi Risk Owner

This identifies which party or individual owns the risk – this is allocated through the contract – the principle being that if the Consultant owns it, it is deemed to have allowed for its price and/or time effect within its tender. If the Client owns it, it will be a compensation event.

vii Mitigation strategy

This registers what actions are proposed and which could be taken to prevent, reduce or transfer the risk. Also, who is responsible for this action and when should the action take place?

viii Allowance in the total of the Prices

How much has the Consultant allowed in its tender for a risk it owns?

ix Programme allowance

How much time has the Consultant allowed in its programme for a risk it owns?

x Client cost allowance

If it is a Client's risk, how much has it allocated to cover a potential compensation event?

xi Risk status

This identifies whether the risk is current, and also whether it is increasing, decreasing or has not changed since it was last reviewed?

xii Last updated

When was an additional risk identified? When was the Early Warning Register last updated?

In effect, the Consultant could seek to qualify its tender by including additional matters within Contract Data Part 2 which it sees or would like to transfer as Client's risks.

Whilst it is not expressly stated who writes the Early Warning Register, it is referred to in Contract Data Part 1, and under Clause 15.4, following an early warning meeting, the Service Manager revises the Early Warning Register to record decisions made and issues the revised Early Warning Register to the Consultant, one would assume that it would be the Client or Service Manager.

Early warning meetings within the Professional Service Contract

The Early Warning Register becomes more important once early warnings lead to early warning meetings, see Clauses 15.2–15.4.

The Service Manager prepares a first Early Warning Register and issues it to the Consultant within one week of the starting date.

The Service Manager also instructs the Consultant to attend a first early warning meeting within two weeks of the starting date, further early warning meetings are held at no longer than the durations stated in the Contract Data, until Completion of the whole of the service.

Early warning meetings can happen in several ways:

- as a standing agenda item in regular (say weekly or monthly) meetings
- as a routine meeting (say once a month)
- ad hoc, when the matter is sufficiently urgent

Meetings can provide a distraction to the "day job", so complying with Clause 15.2 should not place unnecessary demands on the participants. The use of telephone, video or online conferencing works well, particularly with multiple participants in different locations.

Clause 15.2 includes a rare example of an optional action in an NEC contract, when, after the first early warning meeting, either the Service Manager or the Consultant *may* instruct the other to attend an early warning meeting.

Note that the key word in Clause 15.2 is "instruct". If the Consultant calls an early warning meeting, it is *instructing* not merely *requesting* the Service Manager to attend. This clause is clearly intended to promote ownership of the project by the Consultant and the Service Manager, and any issues that could affect it, together with their resolution. Hopefully, in the real world, such meetings are arranged by co-operation and invitation.

Notwithstanding that the Service Manager or the Consultant can instruct the other to attend an early warning meeting, early warning meetings are also held on a periodic basis no longer than the interval stated in the Contract Data until Completion of the whole of the service.

An early warning meeting may only have the Service Manager and the Consultant present, though each may instruct other people to attend if the other agrees, also a Subcontractor may attend; so, in reality, a number of people normally attend the meeting.

On a lighter note, the "rule of meetings" will often apply, in that the productivity of the meeting is often inversely proportional to the number of people who attend it! However, this should not mask the importance of getting the right people to attend.

Service delivery relies on multiple specialists, and these should be encouraged to participate wherever necessary.

Clause 15.3 describes what the attendees at an early warning meeting should do in an effort to either resolve the problem or at least attempt to resolve it:

- making and considering proposals for how the effect each matter in the Early Warning Register can be avoided or reduced,
- seeking solutions that will bring advantage to all those who will be affected,
- deciding on the actions which will be taken and who, in accordance with this contract, will take them,
- deciding which matters can be removed from the Early Warning Register and
- reviewing actions recorded in the Early Warning Register and deciding if different actions need to be taken and who, in accordance with the contract, will take them.

Clearly, the purpose of the early warning meeting is to actively consider ways to avoid or reduce the effect of the matter which has been notified.

In some cases, the matter can be fully resolved, but in others, as the matter may not yet have occurred, it may simply have to be "parked" and recorded as such in the Early Warning Register.

It is the Service Manager's responsibility to revise the Early Warning Register to record the outcome of the meeting and to issue the revised Early Warning Register to the Consultant.

2 The Consultant's main responsibilities

Introduction

The Consultant's main responsibilities are covered within the NEC4 Professional Service Contract by Clauses 20.1–26.2 with further references within the Main Option Clauses.

Providing the Service

Clause 20.1 covers succinctly the Consultant's obligation to Provide the Service in accordance with the Scope.

The Scope is the equivalent of "Works Information" and to a certain degree "Site Information" in an NEC3 Engineering and Construction Contract and will include a detailed description of the service to be provided, possibly including drawings, specifications and schedules, and any constraints in providing the service, for example, working hours, access issues, etc.

It should also include how success or failure in Providing the Service is to be measured.

Note that there is no Site Information within the Professional Service Contract, so any information on the Site (if there is a Site) and surroundings, if relevant, must also be included within the Scope. This could include information on ground investigations, utilities information (electricity, gas, water and telecommunications) regarding existing installations and also environmental issues such as protected species, etc.

In carrying out the service as described within the Scope, the Consultant is obliged to use the skill and care normally used by professionals providing similar services. This is what is commonly known as "reasonable skill and care".

Clause 33.1 requires the Client to provide access to a person, place or thing to the Consultant as stated in Contract Data Part 1 on or before the later of its access date, and the access date for it as shown in the Accepted Programme, and so the Accepted Programme can impose a later date than that shown in the Contract Data, but not an earlier date.

The use of the terms "thing" and "things" is a little curious and usually applies to something that cannot be adequately described! However, it applies to facilities, services, etc., which the Client provides, and should be fully described in the Scope and also specifically referred to in Contract Data Part 1 which identifies what access is to be given and when.

It is critical that the parties understand what is to be provided by the Client, and also it is critical that if there is a charge for the availability and use of those things, it should be clearly detailed within the contract, including how the charge will be measured and when the Consultant has to pay.

The Service Manager may give an instruction to the Consultant that changes the Scope or a Key Date, but after Completion can only give such an instruction if it is necessary to Provide the Service.

The Service Manager may not give an instruction which could require the Consultant to act in a way which was outside his professional code of conduct (Clause 14.4).

Consultants are normally required to adhere to Codes of Conduct which include a duty to behave ethically and are normally set by the professional institution of which they are members.

Examples of codes of conduct normally include obligations to:

- i discharge professional duties with integrity and, in a fair and unbiased matter, to uphold and enhance the standing and reputation of the relevant institution.
- ii only undertake work that they are professionally and technically competent to do and in carrying out that work to adhere to good practice and to demonstrate a level of competence consistent with a professional of their standing. This may also be coupled with the requirement to maintain insurances and indemnify their Clients against risks.
- iii have full regard for the public interest, particularly in relation to matters of health and safety, but also to others' business activities and to the well-being of future generations.
- iv show due regard to the environment and to the sustainable management of natural resources.
- v develop their professional knowledge, skills and competence on a continuing basis and give all reasonable assistance to further the education, training and continuing professional development of others.
- vi not improperly offer or accept gifts or favours which would be interpreted as exerting an influence to obtain preferential treatment.

Examples of a Service Manager giving an instruction which could require the Consultant to act outside his professional code of conduct would be instructing him to do something which was outside his area of expertise, which could cause harm to the Consultant or to others, or to do something which would infringe another's copyright or put him in a position where there would be a conflict of interest.

Fitness for purpose vs reasonable skill and care

In order to consider the responsibilities of the Parties, particularly in respect of professional services, we should consider the principles of "fitness for purpose" and "reasonable skill and care".

If one considers a building project as an example, "fitness for purpose" means producing that building, part of a building or a product fit for its intended purpose.

This is an absolute duty independent of negligence. In the absence of any express terms to the contrary, a Contractor or Consultant who has design responsibility will be required to design "fit for purpose".

"Reasonable skill and care" means performing to the level of an ordinary competent person exercising a particular skill. The criterion is that it is not possible or reasonable to judge someone's professional ability purely by results, for example, a lawyer whose

Client is found guilty is not necessarily a poor lawyer, and a surgeon whose patient dies is not necessarily a poor surgeon. The question to ask is "did that professional apply the correct level of skill and care in carrying out their duties?"

If one then applies the same criterion to a Consultant designer, this is normally achieved by the designer following accepted practice and complying with statutory requirements, codes of practice, etc., but if a new construction technique is involved, the duty can be discharged by taking advice from consultants and warning the Client of any risks involved. This is the "state-of-the-art" defence, meaning that a designer is only expected to design in conformity with the accepted standards of the time.

In the absence of any express terms to the contrary, a Consultant will normally be required to design using "reasonable skill and care".

Design and build contracts will often impose a higher standard than reasonable skill and care on Contractors, i.e. fitness for purpose. This obligation resembles a seller's duty to supply goods that are fit for their intended purpose. The dilemma for design and build Contractors is then that they are bound to a "fitness for purpose" obligation, whereas the Consultant that did that design on his behalf is usually limited to "reasonable and care", a dilemma that continues to haunt design and build Contractors where a design defect arises!

It is worth noting that the "default" position within the Professional Service Contract is to use "the skill and care normally sued by professionals providing services similar to the service".

In the absence of any express terms to the contrary, a construction contract that includes design requires the Contractor to design 'fit for purpose'.

It is also worth noting that the JCT 2016 Design and Build Contract states that:

> the Contractor shall in respect of any inadequacy in such design have the like liability to the Client, whether under statute or otherwise, as would an Architect or, as the case may be, other appropriate professional designer holding himself out as competent to take on work for such design who, acting independently under a separate contract with the Client, has supplied such design for or in connection with works to be carried out and completed by a building contractor who is not the Supplier of the design . . .

This is the JCT version of "reasonable skill and care"!

People

Key persons

The Consultant is required to name his "key persons" within Contract Data Part 2, including their name, job, responsibilities, qualifications and experience.

In Providing the Service, he is then required to employ the same people or a replacement that has been accepted by the Service Manager including the requirement to submit details of the replacement's name, job, responsibilities, qualifications and experience.

The Service Manager may refuse to accept the replacement as their qualifications and experience are not as good as the person they are replacing.

Removing a person

Under Clause 21.2, the Service Manager may instruct the Consultant to remove a person employed by the Consultant. The Service Manager must state his reasons for giving such an instruction.

Many practitioners have queried what sort of reasons a Service Manager could give for removing a person, the typical reasons usually being:

- the person has infringed a health and safety requirement, possibly creating a danger to himself and to others.
- the person is not competent to carry out his duties.
- the person has behaved aggressively towards the Client, the Service Manager or another party.

Clearly, the Service Manager cannot give a reason that is illegal under the law of the contract, for example, instructing the Consultant to remove a person purely on the basis of, for example, their race or gender.

An interesting example used once with the NEC3 Professional Services Contract was that the Employer instructed the Consultant to remove a person as he did not like him, which, unless it is considered as a breach of the obligation under Clause 10 to act in a spirit of mutual trust and co-operation can only be considered as a valid reason!

Working with Others

The Professional Service Contract uses the term "Others" to define parties who are "not the Client, the Service Manager, the Consultant, the Adjudicator or any employee, Subconsultant or supplier of the Consultant", which could include Contractors, other Consultants, utilities companies, third-party regulators, etc.

Clause 22.1 requires the Consultant to "co-operate" with Others in obtaining and providing information which they need in connection with the service.

The Consultant may also have to hold or attend meetings with Others (Clause 22.2), notifying the Service Manager before the meetings in case the Service Manager wishes to attend them.

Subcontracting

The Professional Service Contract has been designed on the assumption that work may be subcontracted with Clause 23 dealing specifically with appointment of Subcontractors. Note also, the Professional Service Contract definition of a Subcontractor under Clause 11.2(14):

> A Subcontractor is a person or organisation who has a contract with the Consultant to provide part of the service, except for the supply of people paid for by the Consultant according to the time they work.

If the Consultant proposes to subcontract work, he is responsible for Providing the Service as if he had not, and also that the contract applies as if a Subcontractor's employees were his own, thus emphasising in both cases that the Consultant is wholly liable to the Client for any Subcontractors he employs.

Clause 23.2 is worth specific mention as the Consultant is obliged to submit the name of each proposed Subcontractor to the Service Manager for acceptance. No Subcontractor can be appointed until the Service Manager has given his acceptance.

Note again that the submission by the Consultant and the acceptance by the Service Manager must both be in a form which can be read, copied and recorded.

As the Consultant cannot appoint the particular Subcontractor until the Service Manager has accepted him, it is vital that the Consultant submits the name of his Subcontractors as early as possible so that the Service Manager has a reasonable period in which to consider them. The Service Manager must accept or give his reasons for not accepting the Subcontractor within the "period for reply" stated in the contract.

A reason for the Service Manager not accepting the Subcontractor is that "his appointment will not allow the Consultant to Provide the Service". This clause probably needs a word of caution to Service Manager who may give the Consultant any one of several reasons for not accepting the Subcontractor, but if the reason is not that "his appointment will not allow the Consultant to Provide the Service" then it is a compensation event under Clause 60.1(8), "the Service Manager withholds acceptance for a reason not stated in this contract".

Examples of non-acceptance of a Subcontractor which could give rise to a compensation event are "I have never heard of your proposed Subcontractor", or "I have heard that he is not doing too well on one of your other sites", neither of which comply with the reason "his appointment will not allow the Consultant to Provide the Service".

Whilst the contract does not state as such, clearly, if the Service Manager has concerns about a Subcontractor, he should raise them with the Consultant who should be expected to provide reasonable substantiation of their competence, possibly in the form of examples of previous work and references. If, despite the Consultant providing this substantiation, the Service Manager refuses to accept the Subcontractor, then it is a compensation event.

Service Managers are often concerned about accepting a particular Subcontractor, probably the best advice being to ask the Consultant for some substantiation and unless the substantiation allays the Service Manager's concerns, to accept the Subcontractor and remind the Consultant that under Clause 23.1, responsibility for that Subconsultant remains with the Consultant.

Whilst the contract does not require the Consultant to use an NEC contract for his Subcontractors under Clause 23.3, he is required to submit the proposed subcontract documents (except pricing information) for his Subcontractors to the Service Manager for acceptance unless an NEC contract is used, or the Service Manager has agreed that no submission is required.

As the Consultant cannot appoint the particular Subconsultant on the proposed subcontract conditions until the Service Manager has accepted them, the Consultant should submit proposed conditions of contract as early as possible so that the Service Manager has a reasonable period in which to consider them.

A reason that the Service Manager can give for not accepting the contract conditions is that:

- they will not allow the Consultant to Provide the Service
- they do not include a statement that the parties to the subcontract shall act in a spirit of mutual trust and co-operation

In addition, Options C and E also require the Consultant to submit the pricing information for each subcontract to the Service Manager for acceptance.

This is an important requirement as costs arising from the subcontracts will become due for payment as Defined Cost within the Professional Service Contract.

3 Time

Introduction

Time, and more specifically, starting date, Completion Date and Key Dates, the Consultant's programme and other time-related issues are covered within the NEC4 Professional Service Contract by Clauses 30.1–35.3, with further references within the Main Option Clauses.

General principles of planning and programme management

A successful construction project, with the associated services, is considered by Clients and Consultants to be one that has been completed on time, to the required quality and also within budget and by Consultants and their Subcontractors as one that has been completed on time, to the required quality and making them a profit.

Clearly, apart from their different perspectives on financial outcome, the objectives of both contracting parties are well matched.

When considering completion on time, various surveys carried out in recent years have shown that only about 33% of construction contracts are completed within the contract period, one-fifth overrun by 40% and a significant number overrun by more than 80%.

There are many factors that can cause delays in construction projects and services:

- i Delays/errors in design, often exacerbated by poor communications between design and construction teams.
- ii Inadequate site investigations, Clients often believe that Consultants designing foundations, etc. should be responsible for assessing ground conditions based on poor and inadequate information provided by them.
- iii Inability to give timely access or possession.
- iv Non-availability of resources, by all parties.
- v Poor performance of parties.
- vi Client changes, compounded by the number and/or timing of these instructions.
- vii Weather and other seasonal problems.
- viii Insolvency of a party.
- ix Unrealistic construction periods set by the Client.

Clearly, matters need to be improved by the parties and between the parties, but many of these factors can be avoided or at least reduced by the parties engaging in proper and effective planning and programming of the works aligned to effective communication between the parties.

Planning and programming

The terms "planning" and "programming" are normally seen as meaning the same thing; however, they are a little different.

What is planning?

Planning may be defined as "the deliberate consideration of all the circumstances concerned with a project in order to evolve the best method of achieving a stated objective". This involves defining the project, setting the overall duration and identifying the risks.

Planning is a vital element in any construction process without which it is impossible to envisage the successful conclusion of any project.

Planning considers:

- what to do
- when to do it
- how to do it
- what resources are required to do it
- who is going to do it
- how long it will take to do it

The motto is

PLAN YOUR WORK AND WORK TO YOUR PLAN!

Planning may be considered on a long or short-term basis, long-term planning including the development of a master plan for the whole project, whilst short-term planning is a much more detailed operation, which considers only the next two to four weeks of operations or a particular element of the work. It is essential that each short-term plan should overlap the other by one to two weeks to allow for past and future activities to be considered.

The planning process considers each operation in turn, analysing the order and sequence of each operation, its dependency on previous operations and the Client's and Consultant's resources to carry it out and the availability of those resources.

It is easier to track progress of operations using short-term rather than long-term planning.

It is considered that there are three planning stages.

Pre-tender planning

This covers the planning considerations during the preparation of a tender and its conversion into a submitted tender price and programme.

The process enables the Consultant to

- establish a realistic price and period on which the tender may be based.

  The Client may set the starting date and the Completion Date for the work or may set the starting date and invite the tendering Consultants to propose a Completion Date.

- consider the most effective economical methods, using the most effective resources.
- assist the build-up and pricing of contract establishment charges and any equipment expenditure.

This is usually carried out in broad outline format as the Consultant may not be successful in his tender.

Typical information that may be gathered and used at the tender planning stage includes:

i a brief description of the service and its location, details of the Client and the various other consultants, key companies, third parties and individuals,
ii details of the local planners and any planning issues or restrictions,
iii access to site for delivery and collection of equipment and materials and also removal of waste (if applicable),
iv details of existing services and the location of new and existing connection points. Details of the relevant service providers,
v general details of the geography and topography of the area, past use, ground investigations, groundwater levels and other local knowledge,
vi location of tips and suppliers including quarries (if applicable),
vii availability of labour, equipment, plant and materials,
viii details of possible Subcontractors and
ix typical weather conditions.

This information is then collected together with information given to the Consultant at tender stage to develop a pre-tender plan. Time-based elements can then be transferred to a pre-tender programme, usually in a bar chart format, which together with the method statements, then aids the estimator in preparing the tender.

Clients should instruct Consultants to submit, at least outline programmes with their tenders, though it must be appreciated that as the Consultant may not be successful with his tender, it will serve only as an indicative outline of how the Consultant would carry out the services should he be successful.

The NEC4 Professional Service Contract provides, as an optional statement, for the Consultant to submit a first programme for acceptance within a specified number of weeks of the Contract Date.

Pre-contract planning

This covers the planning considerations once the contract has been awarded. It is essential that this planning stage takes place prior to commencing the services.

The process enables the Consultant to:

- establish a clear strategy for carrying out the service,
- comply with contract conditions and requirements,
- establish a sequence on which the master programme may be based,
- identify key dates by which parts of the service are required to be completed or the Client is to provide information or resources to enable the Consultant to complete the service to the programme requirements,

- enable the assessment of contract budgets and cash flow forecasts,
- schedule key dates for procurement of key resources and Subconsultant tendering, placing orders and commencement and
- schedule pre-contract meetings.

Project planning

The planning that is required to be implemented in order to maintain control and ensure that the service is completed on time and within the cost limits established at the tender stage.

The process enables the Consultant to

- monitor and update the master programme on a regular basis,
- optimise and continually review resources and
- keep progress under review and report on variances.

What is programming?

Programming is drawing all the planning considerations together into a presentable graphical format such as a bar chart and supporting information and tables that show the various operations, their durations, relationship between the operations, critical path and float and the required resources so that it can be used as a project reporting and control mechanism and form the basis of progress reporting to the Client.

Programme is an essential part of any construction contract, but whilst other contracts often pay only a passing interest to the programme, the NEC family sees it as a vital tool from which to:

i at tender stage, assess the Consultant's ability to carry out the project,
ii measure and report the Consultant's actual progress against that planned,
iii assess the effect of compensation events upon Completion and Price and
iv ensure that the Client takes appropriate action when required.

Preparation of a programme

The process of writing a programme normally consists of three stages:

i Identifying the various services to be carried out

The services may initially be written as a list, identifying each operation in chronological order. That order may depend on natural construction sequence, constraints to access and resource availability.

ii Considering the duration of each part of the service

The duration of each part of the service can only be a forecast rather than an exact science as various issues have to be considered such as length of working days, differing productivity outputs, changes, resource availability, working conditions, changing weather and equipment breakdowns which can affect forecast durations. The Consultant will then make appropriate adjustments as the work progresses.

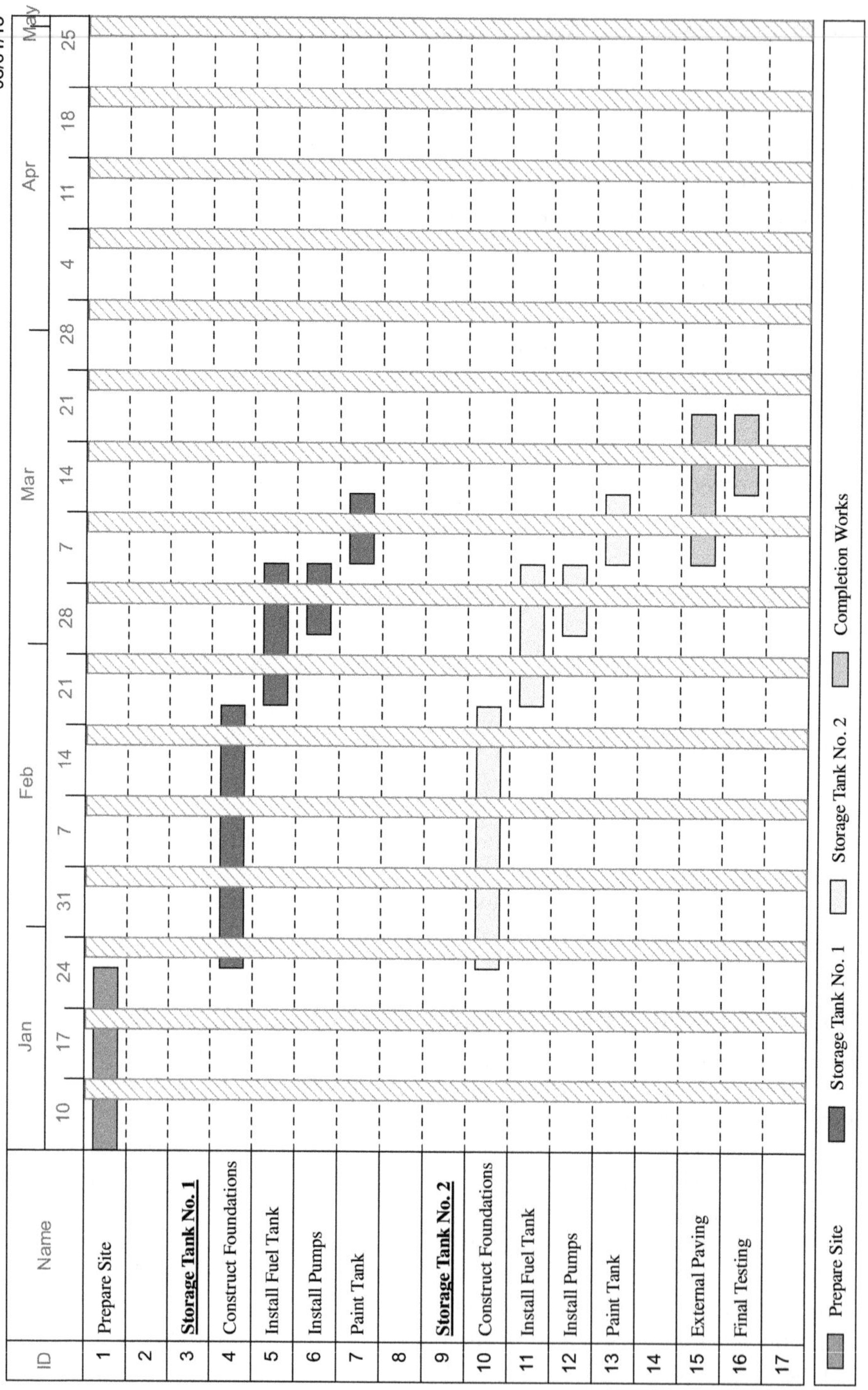

Figure 3.1 "Sample programme without links."

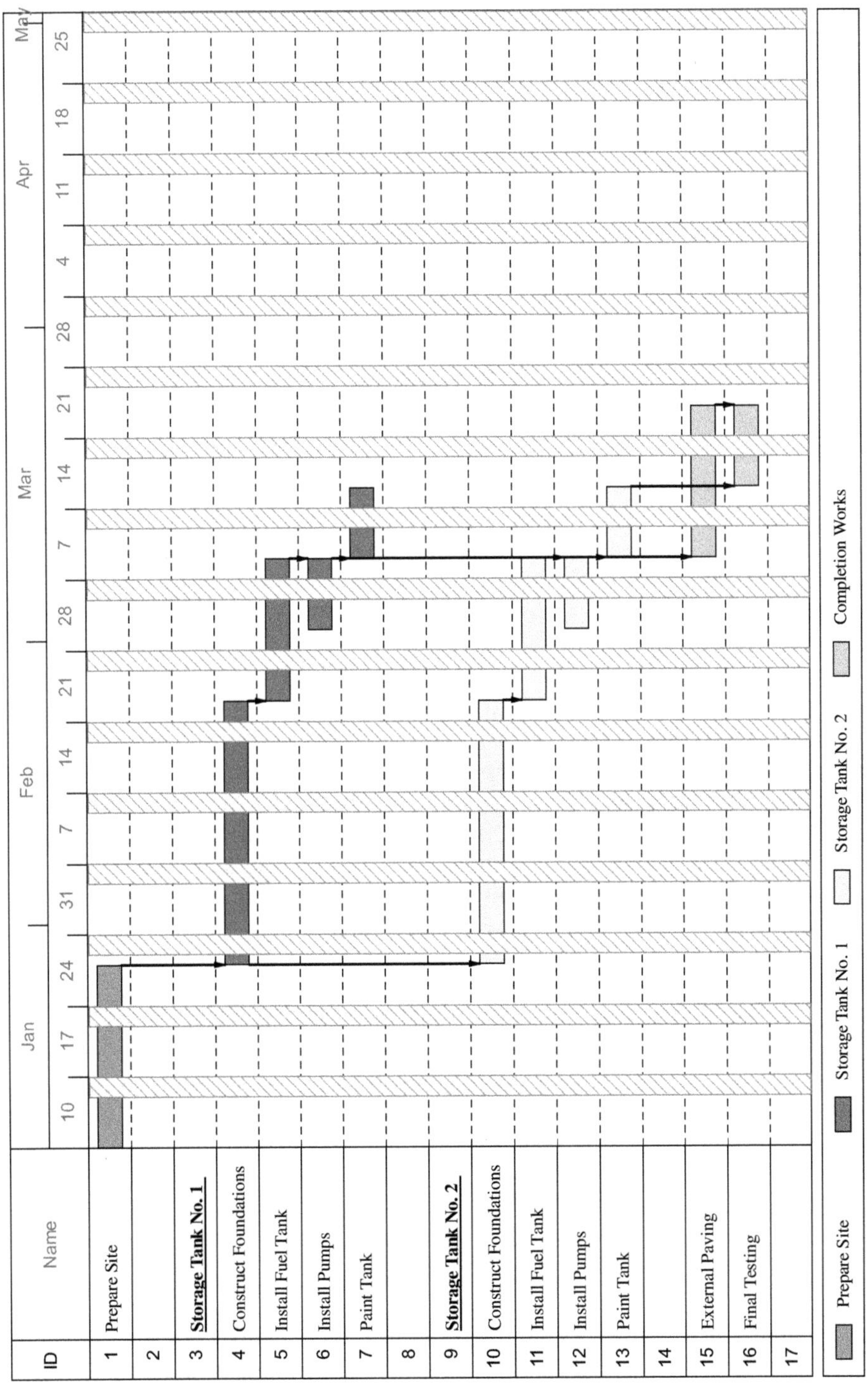

Figure 3.2 "Sample programme with links."

iii The relationship between each operation will involve inserting links between operations that have a dependency on each other, for example, "start to start", "finish to start" and "finish to finish". From this can be established what the logical path is in order to execute the various operations, operations need to start at the same time, what operations have to be completed before another operation can start and what operations must finish at the same time.

Figure 3.1 shows a simple construction programme for the construction of two oil storage tanks. One can interpret from the programme that there is an initial period for preparing the site, the construction of the foundations to both tanks and then commencing once the site has been prepared.

The construction of the tank foundations, the installation of the tanks themselves and the installation of the pumps and tank painting for both tanks then progress concurrently. Whilst the tanks are being painted, the external paving around the tanks can then progress, then the final operation, final testing will progress with completion of this operation coinciding with completion of the external paving and thus completion of the works.

The problem is that without any links, the dependency of the operations upon each other, whilst implied from their positioning within the programme, is not definitive; therefore, if, for example, the construction of the foundations to Tank No. 1 is delayed, either by the Consultant or the Client, the effect on other operations will not be known.

Figure 3.2 includes links that show the logic and dependency of each operation; therefore, if the construction of the foundations to Tank No. 1 is delayed, it is on the critical path, and therefore, the delay to the foundations will have a direct delay effect on all the following operations.

Exactly, the same principles can be applied in constructing a programme for a Consultant to carry out services.

The term programme is not defined in law or within contracts, the dictionary definition normally stating that a programme is "a plan or schedule of activities to be carried out". The Engineering and Construction Contract does not prescribe the form the programme submitted for acceptance must take, so it would normally be a bar chart programme.

Types of Programme

Before considering in detail the requirements of the Professional Service Contract, it is worth reviewing the types of programme in common use, though it has to be borne in mind that when considering professional services, the programme will probably be a lot less complex than a construction programme as there are less factors to consider, so a Gantt Chart will usually suffice.

Bar Charts

Any programme, at any stage of a tender or project, is normally written as a bar chart or Gantt Chart (Henry Laurence Gantt (1861–1919) US Engineer) showing operations which may occur either on or off-site including their sequence and duration, the start and finish of the bar depicting the start and finish date of the operation. Linked bar charts are the usual form of programme in common use on construction contracts. They are often supplemented by other information.

Advantages of bar charts

- The simple format allows them to be readily constructed, and they are user-friendly and understood by most practitioners whether they are Consultants or Contractors.
- They illustrate the specific activities, their durations and during the progress of the work, actual versus planned progress can be clearly shown.
- They can be used at all stages of the planning process through pre-tender, pre-contract and contract planning. They assist in showing the relationship between the pre-tender programme, the master programme and short-term programmes.
- They clearly relate to the sequence of events, and the use of linking on bar charts aids the overlapping of related operations and can enable the critical path to be identified.
- They are easily updated at weekly and monthly time intervals for review purposes and progress reports.
- Key milestone symbols may be introduced to highlight critical dates with regard to key contract stages and information requirements.
- Resources may be readily related to the rate of working.
- The programme forms the basis of financial forecasting for both the Client and the Consultant as a cash flow forecast can be prepared by pricing the programme.

Disadvantages of bar charts

- Without the use of links, they cannot properly depict the relationship and dependency between activities; for example, where an activity must be wholly or partially complete before the next can start.
- Complex interrelationships cannot always be clearly shown.
- They may give too simple a picture, which may be misleading.

Elemental trend analysis (line of balance) charts

The Elemental Trend Analysis (Line of Balance), as with many programming methods originated from the manufacturing industry and whilst little known or used, is beneficial where there is repetition of work either in number of units such as housing or in repetitive floors in constructing or refurbishing multi-storey buildings. The unit or floor numbers are shown on the vertical y-axis and the timeline of the project on the horizontal x-axis.

NB This type of programme is rarely used for professional services.

Advantages of elemental trend analysis charts

- They are very suitable for repetitive work and strict trade sequencing as the flow of work can be clearly demonstrated.
- They are more sophisticated than bar charts and simpler than network diagrams.
- They clearly show the rate of progress of each trade.

Disadvantages of elemental trend analysis charts

They are totally unsuitable for non-repetitive projects and largely unsuitable for limited repetition or for consultancy services.

- They are not easily understood, user-friendly or readily understood by all practitioners.

Project network techniques (precedence diagrams, network analysis, critical path analysis)

Project network techniques comprise a number of methods, but all are related to the dependency and precedence of one activity over another.

Networks may be represented as arrow diagrams where activities are represented by "arrows" and events (start and finishes) by "nodes". Activities take time and are represented as months, weeks, days or even hours, but events do not.

As stated earlier, the bar chart is a clear and simple method of plotting activities, but it has a clear disadvantage in that the relationship between activities cannot be properly represented. With networks, the relationship is immediately apparent, and when a delay occurs, specifically affected activities will stand out.

A project network illustrates the relationships between activities (or tasks) in the project, showing the activities either on arrows or nodes.

NB Again, this type of programme is rarely used for professional services.

Preparing network diagrams

The first operation as with any method is to compile a list of activities to be carried out.

One must then consider against each activity:

- Which activities must be completed before this activity can start?
- Which activities cannot start till this one is completed?

NB: One should always consider when an activity *can* be done rather than when one thinks it *will* be done.

Advantages of network diagrams

- Extremely effective when one is considering highly critical activities with strict dependency of several activities upon each other.
- Float and critical path can be calculated fairly easily from such diagrams.

Disadvantages of network diagrams

- They require a great deal of time and effort to produce them.
- They are not practical for "normal" construction activities, which do not require highly detailed planning and monitoring of progress.
- They do not provide a ready format for financial reporting.

Most major projects are managed using software which combines high quality presentation with ease of revision and updating. This software can also link the master programme to short-term programmes and also resource schedules and cash flow forecasts; however, one must never forget the power of pen and paper.

The traditional approach of contracts

The value of a fully comprehensive programme is often misunderstood and undervalued by construction practitioners; this is often exacerbated by the minimal express requirement for a programme in most standard forms of construction contract, especially for professional services.

Various commonly used forms merely state that the Consultant "submits his master programme for the execution of the services", "states the order in which he proposes to carry out the services" or "shall submit a detailed time programme... and shall submit a revised programme whenever the previous programme is inconsistent with actual progress or with the Consultant's obligations" often without going into any detail as to the required content and structure of the programme, the timing of the submission, the timing of the submission of any revised programme or any formal acceptance of the Consultant's programme.

The danger is that by understating the requirement for a programme within the contract, the true value of a programme is then misunderstood by the parties to that contract, and the programme submission and any acceptance can then descend into a "box ticking" exercise.

Clearly, although the programme requirement within these contracts is not spelt out in detail, it is advisable on any project for the Client to require the Consultant to submit a programme, even though there may be no contractual requirement to do so, and it is also advisable for the Consultant to submit a detailed programme even if there is no contractual requirement to do so. The submitted programme should also identify the requirements to be met by the Client.

The programme must be a dynamic tool that addresses all aspects of the project including design, information and other requirements and the procurement process and lead in times for sub-Consultants and suppliers in addition to the construction operations. It must be flexible and reviewed and, if necessary, revised on a regular basis as the work progresses, and various events and changes come to light including the adjustments to float and time risk allowances.

The NEC Professional Service Contract approach

The NEC Professional Service Contract takes a different approach to other contracts in dealing with programme by clearly stating in detail what the Consultant is required to include in the programme he submits, with sanctions if fails to do so, and also requiring that the Client either accepts the programme or gives his reasons for not accepting it.

The clear programme requirement enables it to be used to assess:

i the Consultant's ability to carry out the project in terms of the intended resources and within the contractual time frame,
ii that all of the contract needs and requirements have been considered,

iii the Consultant's actual progress against that planned and
iv the effect of compensation events upon Completion and Price.

Dates in the contract

There are a number of dates in the Professional Service Contract that require some definitions:

Contract date

The Contract Date (Clause 11.2(4)) is the date the contract came into existence; however, this is achieved, for example, by the Client issuing a letter of acceptance to the Consultant. Note that there is no pro forma contract agreement within the contract, the principle being that parties will enter into contracts in various ways, using their own pro formas, standard letters or exchange of correspondence.

Starting date

In contrast with most forms of contract, the period of time for completion of the service is not stated. Instead, the starting date is decided and defined by the Client in the Contract Data together with the Completion Date. This has the advantage that tenderers know exactly when the services are required to start and to complete.

The starting date and the access dates are stated in the Contract Data. The Consultant can start work from the starting date, but does not have access to a person, place or thing, which can include a Site, until the access date. He also carries risks from the starting date and so has to provide the insurances in the contract prior to that date. The Consultant completes by the Completion Date.

Access date

The date(s), on which the Consultant may have access to and use of various parts of the Site (if required), are stated in the Contract Data.

As stated above, whilst the Consultant can start work on the contract from the starting date, he cannot start work on the Site till the first access date. If the Client has stated different access dates for different parts of the Site, then the Consultant must then programme his activities according to the dates when he will be given access.

The Consultant may not require access on the dates stated in the Contract Data, in which case he should show on his programme the dates when he requires access.

The Client is then required to give access to a person, place or thing by the later of the access date in Contract Data Part 1 and the date for access shown on the Accepted Programme (Clause 33.1).

Completion date

The Consultant is obliged to complete the services on or before the Completion Date as stated in the Contract Data or as may be revised in accordance with the contract, for example, by compensation events.

The Completion Date may be specified by the Client in Part 1 of the Contract Data, or alternatively proposed and tendered by the Consultant in Part 2 of the Contract Data. The inclusion of a definite Completion Date in the contract is essential for proper management of the contract. In many standard consultancy forms, a definite Completion Date is not provided for, or at best, completion is left vague. However, a clear date for completion of the services protects both the Consultant and the Client. The Service Manager is responsible for certifying the date of Completion.

The Service Manager decides the date of Completion and certifies within one week of Completion (Clause 30.2) (Figure 3.3 shows a typical "Completion Certificate" template). Note that the Service Manager is not deciding when the Consultant has to be complete by, but when he has completed.

Failure to complete on time constitutes a breach of the Consultant's obligation. If the Secondary Option X7 (Delay Damages) is included, the Client is then entitled to pre-defined damages for that delay. If X7 is not included, then unliquidated damages will apply.

Whilst many contracts use the terms "Practical Completion" and "Substantial Completion", the Professional Service Contract does not, as these terms are often subject to various interpretations.

The NEC Professional Service Contract is more objective in defining Completion under Clause 11.2(2) as when the Consultant has:

- done all the work which the Scope states is to be done by the Completion Date.

  This first bullet requires the Consultant to comply with the requirements of the Scope and supersedes the traditional terms "practical completion" and "substantial completion".

  This requirement may include not only physical completion of the service but also submission of any relevant documentation, drawings, manuals, training requirements, successful testing requirements, etc.

  By including these requirements within the Scope, Completion has not been achieved until they are completed.
- corrected Defects which would have prevented the Client from using the service and Others from doing their work.

  The service may contain Defects, though these may not be significant enough to prevent the Client from practically and safely using the service. If Defects exist, Completion may be certified with the requirement that the Consultant corrects the Defect before the end of the defect correction period following Completion.

There is also a fall back within Clause 11.2(2) in that if the work which the Consultant is to do by the Completion Date is not stated in the Scope, Completion is "when the Consultant has done all the work necessary for the Client to use the services and for Others to do their work". This broadly aligns with the traditional terms "Practical Completion" and "Substantial Completion", where the Client is able to make beneficial use of the services.

The Service Manager is responsible for certifying Completion, as defined in Clause 11.2(2), within one week of Completion (Clause 30.2).

Normally, the Consultant will request the certificate as soon as he considers he is entitled to it, but such a request is not essential.

Contract:	COMPLETION CERTIFICATE
Contract No:	Certificate No................................

Completion achieved on: .... (Date)

The Completion Date is: .... (Date)

The defects date is: .... (Date)

The defects on the attached schedule are to be corrected by: .... (Date)

The outstanding works on the attached schedule are to be completed by: .... (Date)

Comments:

Signed: ....(Service Manager) Date: ....

Copied to:

Service Manager ☐ Consultant ☐ Other .... ☐ File ☐

Figure 3.3 Sample completion certificate.

Key date

This is defined in Clause 11.2(9) as the date by which work is to meet a stated Condition. The Consultant does the work so that the Condition stated for each Key Date is met by the Key Date.

It is an optional clause allowing the Client, should he wish, to include dates when the Consultant must complete certain items of work, possibly to allow others, which could be the Consultant or another consultant to carry out other work.

If used, the Condition and the applicable Key Date are defined by the Client in Contract Data Part 1. Note that there is no provision for adding Key Dates; the Service Manager may change a Key Date, but it is a compensation event (Clause 60.1(4)).

"Key Dates" must be differentiated from "Completion" or "Sectional Completion", in that with Key Dates, the Client does not "take over" the services, whereas with Completion and Sectional Completion, he does. In this respect, there are no pre-determined delay damages applied to the Consultant who does not meet a Condition by a Key Date, though there is provision for the Client to recover cost due to that failure.

Under Clause 22.3, if the Service Manager decides that the Consultant has not met the Condition stated by the Key Date, and the Client incurs additional cost in carrying out work or paying an additional amount to Others in carrying out work on the same project as a result of that failure, then the Consultant will be liable to pay that amount. One could question whether, if the Client has to purchase something, for example, he has to buy an item or document which the Consultant was supposed to buy and submit to him, whether that is classified as "carrying out work"?

Note also that "the Service Manager decides"; therefore, he makes the unilateral decision to decide what the Consultant owes him and is not expressly required to provide details of how that decision has been calculated. Clearly, the Consultant can refer a decision to adjudication if he feels that it is either unreasonable or has been calculated wrongly.

By referring to "the same project", the Client cannot claim costs incurred on another project as a result of the failure of the Consultant to meet a Key Date on "this project". This cost is assessed by the Service Manager within four weeks of when the Consultant actually does meet the Condition for the Key Date.

Example

The Consultant has been appointed under the NEC4 Professional Service Contract to provide Scope for repairs to an existing drainage system, which is being repaired by a Contractor under the NEC4 Engineering and Construction Contract.

The works are to be divided into ten phases, the arrangement being that the Consultant provides the Scope for a phase, and then the Contractor carries out the necessary repairs to that phase whilst the Consultant prepares the Scope for the next phase.

Key Dates have been inserted into the Professional Service Contract for Scope for successive phases to be available to meet the Contractor's programme for the whole of the works. The necessary work to be carried out by the Consultant in preparing the Scope for each phase for release to the Contractor will be the Condition, and this will have a Key Date attached to it.

If the Consultant then fails to meet the Condition by the Key Date and the Client incurs a cost in having to postpone the Contractor's progress, it is a compensation event for the Contractor under Engineering and Construction Contract Clause 60.1(3), and the cost of which is then recovered from the Consultant under Professional Service Contract Clause 22.3.

The cost must be incurred on the same project, so, for example, if the late release of the Scope causes the Client to incur a cost on another project, then this is not paid by the Consultant.

The Consultant's programme

The Professional Service Contract has quite comprehensive and specific requirements for a programme to be submitted (see Figure 3.4) by the Consultant, these requirements being defined by nine bullet points within Clause 31.2.

Some Consultants say that the requirement within Clause 31.2 is very onerous and, in some cases, excessive; however, it is critical that the programme shows in a clear and transparent fashion what the Consultant is planning to do, when and how long each operation will last and what he needs from the Client, the Service Manager and others to be able to comply with it.

There is nothing within the requirements of Clause 31.2 that a competent Consultant would not ordinarily include within a professionally constructed programme submitted for the benefit of himself and the receiving party. Whether and how he chose to show the information could be another matter.

This information provides both parties with a sound means of assessing the impact of change on the contract, minimising the need for conjecture leading to possible disputes. The information required includes not only dates for various activities and operations but also method statements and other information stated in the Scope.

There are two alternatives for the Consultant's submission of his first programme for acceptance:

i He may submit, or be required to submit, a programme with his tender in which case it is referenced by the Consultant in Contract Data Part 2. Note that under Clause 11.2(1), "the Accepted Programme is the programme identified in the Contract Data or is the latest programme accepted by the Service Manager. The latest programme accepted by the Service Manager supersedes previous Accepted Programmes".

Therefore, any programme referred to in Contract Data Part 2 automatically becomes the first Accepted Programme even though it may not necessarily comply fully with the requirements of Clause 31.2.

ii If a programme is not identified in Contract Data Part 2, the Consultant submits a first programme to the Service Manager for acceptance within the period of time after the Contract Date, this period of time being stated in Contract Data Part 1.

There is no default position, found in some other contracts, where no response to a programme submission is deemed as acceptance of it.

Within the NEC4 Engineering and Construction Contract, Clause 50.3 provides (and has always provided) that if no programme is identified in Contract Data Part 2, one-quarter of the Price for Work Done to Date is retained until the Contractor has submitted the first programme showing the information the contract requires.

Contract: .... Contract No:	**CONSULTANT SUBMISSION** C/S No.......................................
Section A: To: Service Manager	Submission of: ☐ Drawings ☐ Programme ☐ Test Results ☐ Other

The following information is transmitted for your acceptance:-

Description	Date	No./Rev.	Copies

Response reqd. by ....

Signed: (Consultant) .... Date ....

Section B: Reply

To: Consultant

The submission is returned:-

☐ Accepted	☐ Accepted as noted
☐ To revise and resubmit	☐ Rejected as noted

Notes:

Signed: (Service Manager) .... Date:...............

Copied to:

Service Manager ☐ Consultant ☐ Other .... ☐ File ☐

Figure 3.4 Sample consultant submission form.

This has never been included within the Professional Service Contract, but a number of Clients have added it as a Z clause, particularly where time and the need for a detailed programme submitted early are of the essence.

This provision has now been added into the NEC4 Professional Service Contract as Clause 50.5, whereby one-quarter of the Price for Service Provided to Date is retained until the Consultant has submitted the first programme showing the information the contract requires.

Clause 31.2

The *Consultant* shows on each programme that he submits for acceptance:

- The starting date, access dates, Key Dates and Completion Date

  These dates are stated by the Client in Contract Data Part 1. With regard to the Completion Date, if the Client has decided the Completion Date for the whole of the services, then it is inserted into Contract Data Part 1. Alternatively, if the Consultant is to decide the Completion Date, he inserts this into Contract Data Part 2 as part of his tender.
- Planned Completion

  This is the date the Consultant *plans* to complete the service; this is different to the date by which the Consultant is *required* to complete the service as stated in Contract Data Part 1 or as may be revised in accordance with the contract.

  Although the Consultant can complete any time between his planned Completion date and the Completion Date, failure to complete on or before the Completion Date constitutes a breach of the Consultant's obligation, and if Option X7 has been selected, the Consultant is liable to pay the Client delay damages.

  The difference between the dates when the Consultant *plans* to complete and when he is *required* to complete is float within the programme. This float, often referred to as "terminal" float, belongs to the Consultant and is attached to the whole programme. If a compensation event occurs, a delay to the Completion Date is assessed as the length of time that planned Completion is later than planned Completion as shown on the Accepted Programme (Clause 63.5).

  It is vital that, as planned Completion can change on a regular basis, the Consultant shows a realistic planned Completion in each revised programme submitted for acceptance, but also that the Consultant and Service Manager comply with the time scales in the contract for assessing and implementing compensation events as it is not unusual for planned Completion to be shown later than the Completion Date leading one to believe the Consultant is in delay, yet the reason for this is either that the time scale in the contract is being adhered to but is giving a false picture as a programme is submitted whilst the Consultant is still preparing his quotation for a compensation event, or that compensation events have not been assessed and implemented in accordance with the time scales stated in the contract.
- The order and timing of the operations which the Consultant plans to do in order to Provide the Service

  This shows the sequence, duration and dates of the various operations the Consultant plans to do and should include dependencies and links between the various operations. It should also show sequence, duration and dates in the procurement process, finalisation of information, placing orders, etc. It should also show work to be carried out off site.

- The order and timing of the work of the Client and Others as last agreed with them by the Consultant or, if not so agreed, as stated in the Scope

  The Consultant will state what he requires the Client and Others to do in order that the Accepted Programme can be met. These dates may already have been agreed with the Consultant, but may also have been inserted into the Scope by the Client. "Others" are parties outside the contract i.e. not the Client, the Consultant, a Subconsultant or the Adjudicator. Examples of "Others" would be planning authorities, utilities companies, etc.

  This requirement is often provided in the form of an "Information Required" spreadsheet identifying what information is required and the latest date for providing it, failing which a compensation event can occur (Clause 60.1(3)). It is important to recognise that the programme is not just the bar chart but may include spreadsheets, schedules and graphs.
- The dates when the Consultant plans to meet each Condition stated for the Key Dates and to complete other work needed to allow the Client and Others to do their work.

  If there are Key Dates in the Contract (Clause 11.2(9)), then the Consultant must show how and when he intends to meet each Condition by these Key Dates. It is curious that the Consultant has to show "how" he intends to meet a Key Date, though his programme may show operations related or preceding it; but in most cases, the Consultant simply shows on his programme that he plans to meet the Condition by the Key Date.

Provisions for

- Float

  Float is any spare time within the Consultant's programme, after time risk allowances have been included, and represents the amount of time that operations may be delayed without delaying following operations (free float) and/or planned Completion (total float). It can also represent the time between when the Consultant plans to complete and when the contract requires him to complete (terminal float).

  Float absorbs to a certain extent the Consultant's own delays or the delays caused by a compensation event, thereby lessening or avoiding any delay to planned Completion. In effect, no delay arises unless float on the relevant and critical operations reduces to below zero.

  Programming is never an exact science, so float gives some flexibility to the Consultant in respect of incorrect forecasts or his own inefficiencies. As the work progresses, the float will change as output rates change, Consultant's risk events take place and also compensation events arise.

  The general belief, certainly amongst many Consultants, is that float belongs to them as they wrote the programme and therefore have the right to work to that programme and use any float that it contains. Conversely, Clients believe that if it is free time, then they have the right to use it, but it actually depends where the float is and what it is for.

  In principle, float other than terminal float or time risk allowances is a shared resource, it is spare time, it belongs to the project and therefore may be used by whichever party needs it first.

  There are three primary types of float:

  i The amount of time that an operation can be delayed before it delays the earliest start of following operation (free float).

ii The amount of time that an operation can be delayed before it delays the earliest completion of the services (total float).
iii The amount of time between planned Completion and the Completion Date (terminal float).

Generally, with other contracts, if the Consultant shows that he plans to complete a project early, then he is prevented from completing as early as he planned, but he still completes before the contract Completion Date, then although there may be an entitlement to an extension of time under the contract, for example, exceptionally adverse weather, none will be awarded as there is no delay to contract completion, but he may have a right to financial recovery subject to him proving loss and/or expense.

The Professional Service Contract deals with this in a different way. If the Consultant shows on his programme planned Completion earlier than the Completion Date and he is prevented from completing by the planned Completion Date by a compensation event, then, when assessing the compensation event, Clause 63.5 states "a delay to the Completion Date is assessed as the length of time that, due to a compensation event, planned Completion is later than planned Completion as shown on the Accepted Programme"; therefore, any terminal float is retained by the Consultant, the period of delay being added to the Completion Date to determine the change to the Completion Date.

Any delay to planned Completion due to a compensation event therefore results in the same delay to the Completion Date. Therefore, the extension is granted on the basis of time the Consultant is delayed i.e. entitlement, not on how long he needs to achieve the current Completion Date.

- Time risk allowances

  Time risk allowances are essentially a form of float and are included within the duration of a specific operation. Essentially, a time risk allowance is the difference between what is the most "optimistic" (shortest) duration to complete an operation and what is a "realistic" (expected) duration bearing in mind the risks that the Consultant may face in completing that operation. It is important that the Consultant's staff are fully informed as to how the time-risk allowances are being presented, so there is no misunderstanding of how the works have been programmed and priced.

  The Consultant's time risk allowances are to be shown on his programme, but there are no requirements in the contract as to how these allowances should be shown, three possible methods being:

i Show a single bar for an operation, within which the time risk allowance is shown. This is difficult to do with some planning software.
ii Show a separate bar which represents the time risk allowance for a specific operation. This however adds to the lines on the programme.
iii Show a separate column identifying the time risk allowance included within the total duration of the operation.

Clause 63.8 states that the assessment of the effect of a compensation event includes risk allowances for cost and time for matters that have a significant chance of occurring and are at the Consultant's risk.

It follows that they should be retained in the assessment of any delay to planned Completion due to the effect of a compensation event. These allowances are owned by the Consultant as part of his realistic planning to cover his risks in pricing his tender and also in managing changes. They should be either clearly identified as such in the programme or included in the time periods allocated to specific activities.

Example of a time risk allowance

The Consultant has an operation in his programme for "carrying out survey". He believes that if he is able to make regular progress, there are no staff absences, the weather is reasonable and there are no problems with the equipment he is proposing to use, then the survey will take him three weeks to complete.

However, his knowledge and experience tell him that rarely is everything "reasonable" and almost inevitably he will encounter some delay, so a realistic time scale would be four weeks. He should then show a four-week duration for the survey, within which one week is his time risk allowance.

It is important to recognise that time risk allowances belong to the Consultant and are retained in the assessment of any delay to planned Completion due to a compensation event.

- Access to a person, place or thing if later than its access date
- Information and things to be provided by the Client
- Information and approval from Others
- Health and safety requirements
  This will include time taken in inducting new staff and operatives.
- The procedures set out in this contract
  This can include time scales for acceptance of design, etc.
- The dates when, in order to Provide the Service in accordance with his programme, the Consultant will need

Generally, the Consultant should be given access to parts of the Site (if required), acceptance of his design, Subcontractors and programme in a timely manner to allow him to Provide the Works.

However, in respect of access, the Consultant may be in delay and may therefore not require access on the date shown in the contract, and there may also be certain operations for which there may be a long lead in time, which the Client or Others may not appreciate, and therefore the Consultant can include the requirement for acceptances and information at an early date.

This information may be highlighted as a milestone on the bar chart itself or attached to it in the form of information required schedules, resource schedules, etc.

Anything submitted by the Consultant as part of his programme must clearly and simply show the relevant operations for which access, plant and materials or information are required, and the implications if they are not provided.

It is important that the Consultant and the Service Manager then manages the process by raising any concerns in respect of their provision as early warnings.

- For each operation, a statement of how the Consultant plans to do the work identifying the resources which will be used

The Consultant is required to provide a statement of how he intends to carry out each and every item on each programme he submits for acceptance.

Whilst accepting that the contract requires this information, this can be a somewhat idealistic clause that requires the Consultant to comply with the daunting and in many cases somewhat unreasonable requirement to produce all the information in his first programme submitted for acceptance, which particularly on a large complex project with a long duration, the Consultant may not yet know how he will carry out each operation and what resources he will use. He also may not have appointed a Subconsultant for a part of the work so will not have method statements from the company who is actually carrying out the works.

It is also a significant task for the Service Manager to accept all the information the Consultant has submitted, so it may be appropriate in this respect for the Consultant and the Service Manager to agree a rolling programme of submission of method statements; as the work progresses, the methods and resources may become clearer, perhaps each submission covering the next three or four months' operations.

Method statements have become increasingly important in the construction industry particularly as health and safety legislation makes the parties more responsible for stating the methodology they are going to use in carrying out the works, and how they have provided for the various risks involved. For a programme to be meaningful, it should always be linked to method statements so that the methodology if carrying out the works and also the resources the Consultant intends to use are clearly stated.

The contract does not specify the form these statements should take, and in many cases, a Consultant will spend a great deal of time and paper producing detailed statements, which can be almost meaningless. They may be presented as a written statement or in tabular form, but they should clearly show how the Consultant intends to do the work, main headings being:

- i Title of each operation
- ii Description of method
- iii Quantities (where relevant)
- iv Principal Equipment to be used with outputs and durations
- v Labour type and gang sizes with outputs and durations
- vi Plant to be installed
- vii Materials to be used

NB: Cost is not normally included in method statements.

When preparing method statements, the Consultant should always consider and show:

- i What is to be done
- ii How it is going to be done
- iii Who is going to do it
- iv Where it is to be done
- v When it is to be done

This information will then initially allow the Client to consider whether the Consultant is providing sufficient resources to Provide the Works within the time available and also safely, but also as the programme is revised and methods of construction and resources used may change, the Client is then kept up to date as to the Consultant's intentions.

This information will also prove invaluable when assessing compensation events.

- Other information which the Scope requires the Consultant to show on a programme submitted for acceptance.

Again, this shows that the Scope is not just the drawings and specifications, but information on all that the Consultant is required to do in Providing the Service.

From these requirements, one can see that the programme under a Professional Service Contract is not just the bar chart, but a collection of documents including method statements, risk assessments, resource analyses, etc. All of these are submitted by the Consultant as his programme, and all of these must be considered by the Service Manager in determining whether to accept the programme. Extreme care should be taken to only require documentation which has a purpose and use in the administration.

Note that the Scope may also define the format of a programme to be submitted by the Consultant.

Activity schedule-based options A and C

Clause 31.4 of Options A and C requires that the Consultant provides information which shows how each activity on the Activity Schedule relates to the operations on each programme which he submits for acceptance.

Clause 55.3 of Options A and C requires that if the Consultant changes his method of working so that the activities on the activity schedule do not relate to the operations on the Accepted Programme, and he submits a revision of the activity schedule to the Service Manager for acceptance.

Under Clause 55.4, a reason for not accepting a revision to the Activity Schedule is that

- it does not comply with the Accepted Programme,
- any changed Prices are not reasonably distributed between the activities or
- the total of the Prices is changed.

The Service Manager's response

Within two weeks of the Consultant submitting a programme to him for acceptance, the Service Manager either accepts the programme or notifies the Consultant of his reasons for not accepting it.

A reason for not accepting a programme is as follows:

- The Consultant's plans which it shows are not practicable.

The Consultant could be planning to use a particular working method or piece of Equipment that the Service Manager believes would not be appropriate? Or the Consultant has allowed a particular quantity of labour or a planned output, which the Service Manager believes is not practicable.

- It does not show the information which the contract requires.

It may not show Key Dates or Access Dates, method statements or any indication of float.

- It does not represent the Consultant's plans realistically.

The Service Manager may not believe that the Consultant can complete an operation within the time he has allowed, or he could be showing planned Completion, which the Service Manager believes is too early and therefore unrealistic.

- It does not comply with the Scope.

The Consultant may not have allowed for a constraint within the Scope. Or the programme may simply be in the wrong format. For example, the Scope may prescribe that the programme has to be compiled and submitted using a particular brand of software, but the Consultant has used a different brand.

It has to be said that in many cases, the Service Manager does not accept because the Consultant has misunderstood the contract requirements, he has not included all the stated constraints or has not taken proper account of all the factors included in the Scope.

It is important that the Service Manager either accepts or does not accept the Consultant's programme, and if he does not accept it, he clearly states why he does not accept it.

There is no default position if the Service Manager fails to accept or reject the Consultant's first programme or subsequent submitted programmes.

If he does not accept the programme, and it is not for a reason stated in the contract in Clause 31.3, then it is a compensation event (Clause 60.1(8)), though the Consultant is not prevented from commencing the services, so it is difficult to envisage what the Consultant would be seeking to recover from the breach through the compensation event process?

It has been known for Service Managers (formerly Employers in the NEC3 Professional Services Contract) to respond to the submission of the programme either by making no comment at all within the two weeks they have to respond or by stating that the programme is accepted on condition that the Consultant makes certain amendments to it.

Because of these issues of non-response, or at least vague responses, the NEC4 Professional Service Contract now has a new provision within Clause 31.3 in that, if the Service Manager fails to reply to the Consultant's submission within the time allowed, the Consultant may notify to that effect.

If the failure to respond continues for a further one week after the Consultant's notification, it is treated as acceptance by the Service Manager. This will hopefully goad an "errant" Service Manager to do what the contract requires it to do.

Note that the acceptance of a programme that shows the Consultant completing late whether deliberately or via the new provision does not mean that the Service Manager accepts the delay; it merely acknowledges that the delay is reasonably reflected within the programme, and if Option X7 is selected, the Consultant is still liable for delay damages.

The Accepted Programme:

- effectively provides an agreed record of the progress of the service and where the delays have come from,
- provides a realistic base for future planning by both the Consultant and Service Manager,
- is the base from which changes to the Completion Date are calculated and
- is the base from which additional costs are derived.

Because each operation has a method statement and resources attached to it, the change in resources can be calculated and hence the change in costs. Because information is so much more transparent, there is more scope for working together.

Revised programmes

The timing for submission of revised programmes for acceptance is defined by Clause 32.2:

- within the period for reply after the Service Manager has instructed the Consultant to
  For example, the Service Manager could instruct the Consultant to submit a programme to enable an issue to be discussed in an early warning meeting?
- when the Consultant chooses to
  Again, the Consultant may feel it beneficial to submit a programme to be discussed in an early warning meeting
- and in any case, at intervals no longer than that stated in Contract Data Part 1.
  This is the longest period for submission of a revised programme.

The revised programme shows the following:

- The actual progress achieved on each operation and its effect upon timing of the remaining work

The progress achieved may be shown on the programme itself or appended to it in the form of a progress report. Operations which are not yet completed will normally be shown as percentage complete, being the progress achieved to date against the total work within the operation on the date the assessment is made. It is useful if the Consultant and the Service Manager agree the progress statement before the revised programme comes into being. The submitted revised programme is then an undisputed statement of fact.

The revised programme should reflect planned versus actual progress including early or delayed start, early or delayed completion of each operation.

It is important that, if the Consultant states a percentage completed to date, he should show actual progress achieved rather than time expired to date. Also, the Consultant should not just make statements such as "operation on programme", as this may be misleading. An operation on programme does not necessarily mean that the operation is not delayed, for example, the Consultant is expecting to complete an operation late, then stating the words "operation on programme" can be interpreted as either in accordance with the contract or progressing late in accordance with his expectations.

- How the Consultant plans to deal with any delays and to correct notified Defects

The Consultant should show in his revised programme any delays which may or may not be caused by him and also time needed to correct his own Defects.

- Any other changes which the Consultant proposes to make to the Accepted Programme

An example of this could be that the Consultant resequences part of the work or proposes to use a different method or different equipment to that specified by him in a previous method statement.

It is therefore very much a "living" programme. If the Consultant does not submit a programme which the contract requires, then the Service Manager can assess compensation events, without receiving a quotation from the Consultant (Clauses 64.1 and 64.2).

Note that the NEC3 Professional Service Contract referred to "the effects of implemented compensation events and of notified early warning matters" then in a later revision to "the effects of implemented compensation events", but this has now been omitted from the NEC4 Professional Service Contract.

It was curious that the effects of notified early warning matters were no longer required to be shown on revised programmes? If an early warning had been notified in accordance with Clause 15.1, there was a fair possibility that it may have an effect on the programme.

Note also that a compensation event is not "implemented" until the quotation has been accepted or assessed by the Service Manager. Whilst the Completion Date is not changed until a compensation event has been implemented, there is a danger in only showing compensation events that have been implemented as the revised programme is not reflecting reality.

Record keeping

One of the most important, but often understated elements of managing a construction contract and the various services, whether one is a Consultant or a Client, is the keeping of concise and timely records. It is beneficial if the records are agreed by the Consultant and the Service Manager. These are then undisputed records of fact; it is only the interpretation of these facts that might subsequently be at issue.

A good record keeping system will enable the Consultant or the Service Manager to notify early warnings promptly, assist the preparation of programmes, enable payments to be made under Options C to E and also prove invaluable in pricing and assessing compensation events, all of which have time scales applied to them, so readily available records are a must.

Also, on rare occasions, when a dispute arises and is referred to adjudication, the tribunal records are invaluable.

Records can include:

- Diaries
- Meeting minutes
- Photographs including dates
- Resource returns
- Weather records
- Correspondence
- Financial accounts and records

Instructions to stop or not to start work

The Service Manager has the authority to give instructions to the Consultant to stop, not to start work for any reason and may later instruct him to re-start it. The instruction would be a compensation event Clause 60.1(4).

However, if the instruction was required due to a fault of the Consultant, for example, unsafe working, the Service Manager should instruct the Consultant that the Prices, the Completion Date and the Key Dates are not to be changed.

Acceleration

Acceleration in many contracts normally means increasing resources, working faster or working longer hours so that completion can be achieved by the Completion Date, in effect the Consultant is catching up to recover a delay. However, within Clause 35 of the NEC4 Professional Service Contract, acceleration means increasing resources, working faster or working longer hours so that Completion can be achieved *before* the Completion Date.

The Service Manager may instruct the Consultant to submit a quotation for acceleration. He also states and changes to the Completion Date to be included within the quotation. The quotation must include proposed changes to the Prices and a programme showing the earlier Completion Date and Key Date. As with quotations for compensation events, the Consultant must submit details within his quotation; however, the contract does not prescribe how the quotation is to be priced, for example, based on the forecast Defined Cost plus the percentage for Fee.

The Consultant may give his reasons for not submitting a quotation within the period for reply; there is nothing the Service Manager can do if the Consultant does not submit a quotation or if the Consultant's quotation is not accepted by the Service Manager in that he could make his own assessment as he would with a compensation event.

Acceleration can only be undertaken by agreement between the Client and the Consultant and cannot be imposed on the Consultant or any assessment imposed upon him without his agreement.

When the Service Manager has accepted the quotation for acceleration, he changes the Prices, Completion Date and Key Dates and accepts the revised programme.

If Option X7 is selected, any delay damages become payable by the Consultant in the normal way from the revised Completion Date.

4 Quality management

Introduction

Quality management is covered within the NEC4 Professional Service Contract by Clauses 40.1–42.2.

Quality

Quality and its management is a major management function within the construction industry, including Consultants.

Unless a consulting organisation can guarantee its Client a quality product, it cannot compete effectively with others in the modern market.

Quality often stands alongside cost as a major factor in Consultant selection by Clients. To be competitive and to sustain good business prospects, quality systems need to be designed and maintained which are efficient and evidential.

The role of quality management for a consulting organisation is not an isolated activity, but intertwined with all the operational and managerial processes of that company.

The modern concept of quality is considered to have evolved over many years through the principles of quality control and quality assurance.

Quality

The term 'quality' is often used to describe prestige products such as expensive jewellery and motor cars. However, although applicable to these items, the term 'quality' does not necessarily refer to prestigious products but merely to the fitness of the product to the customer's requirements. Quality is therefore described as meeting the requirements of the customer.

Quality control

Quality control introduces inspection to stages in providing services, ensuring that they are undertaken to specified requirements. The quality is measured by comparing the service actually carried out against the sample initially provided.

Inspection is the process of checking that what is produced is what is required. It is about the identification and early correction of Defects.

The term 'quality control' is defined by an interpretation of its elements: 'quality' and 'control'.

Control

The concept of being 'in control', or having something 'under control', is readily understood – we mean that we know what we intend to happen and are confident that we can ensure that it does.

Quality control is primarily concerned with Defect detection and correction. The main quality control technique is that of inspection and statistical quality control techniques (i.e., sampling) to ensure that the work produced and the materials used are within the tolerances specified. Some of these limits are left to the inspector's judgement, and this can be a source of difficulty.

The major objectives of quality control can be defined as follows:

- To ensure that the completed work meets the specification
- To reduce customers' complaints
- To improve the reliability of products or work produced
- To increase customers' confidence
- To reduce production costs

Quality assurance

Quality assurance is the process of ensuring that standards are consistently met thereby *preventing* Defects from occurring in the first place.

"Fit for purpose" and "right first time, every time" are the principles of quality assurance.

Quality assurance concentrates on the production or construction management methods and procedural approaches to ensure that quality is built into the production system.

Quality assurance is described under the following headings:

- Evolution of quality assurance from quality control
- Definition of quality terms
- Quality assurance standards
- Developing and implementing a QA system

Defining a Defect

A Defect is defined under Clause 11.2(6) of the NEC4 Professional Service Contract as "a part of the service which is not in accordance with the Scope or the applicable law".

Note that it is considerably more difficult to identify a Defect in a Consultant's service than it is within a Contractor's works, also the Defect may not become apparent for some time.

Note that within the definition above, a Consultant failing to exercise reasonable skill and care in carrying out the services is not defined as a Defect.

The Scope provides the reference point for what the Consultant has to do to Provide the Service. If a part of the service provided by the Consultant does not comply with the Scope, then the Consultant is obliged to correct the Defect so that it does comply with the Scope.

In most cases, the work done will fall short of the Scope, but as the words "not in accordance with" are used, work which exceeds the Scope would also be a Defect. Clearly, the Service Manager is likely to accept such a Defect!

A Defect may not necessarily mean that the services provided are not fit for their intended purpose, just that they are not in accordance with the Scope.

It is also possible that the services provided are "defective" but as they comply with the Scope, there is no Defect.

Note that there is a new clause (Clause 42) whereby the Consultant and the Service Manager could each propose to the other that the Scope be changed so that a Defect does not have to be corrected, and in return, the Consultant could submit a quotation for reduced Prices, an earlier Completion Date, or both.

Limitation periods

Whilst the Professional Service Contract has a period from Completion to the Defects Date during which the Consultant is liable for correcting Defects, most countries also have a statutory limitation period for latent Defects which come to light at a later date, beyond which a claim cannot be made for defective work.

In the UK, for example, the time limit for actions founded on simple contract is six years, and for a specialty contract or deed, the time limit is twelve years; in both cases, this is measured from the date on which the cause of action accrued; normally, this is taken as the date of Completion.

Quality management system

Clause 40.1 requires the Consultant to operate a quality management system as stated in the Scope and complying with the requirements stated in the Scope.

A quality management system is a set of interrelated or interacting elements that companies and other organisations use to direct, manage and control how quality policies are implemented and quality objectives are achieved.

A process-based quality management system uses a process approach to manage and control how its quality policy is implemented and quality objectives are achieved.

The quality management system will normally be to ISO9000 standards, ISO9000 defining a Quality Management System as "co-ordinated activities to direct and control an organization with regard to the degree to which a set of inherent characteristics fulfills the requirements".

ISO 9001 sets out the basic requirements for a quality management system, generally that it enables a company to consistently provide products or services that enhance customer satisfaction whilst meeting applicable statutory and regulatory requirements. It is essentially about providing quality assurance to customers.

It can be used by any organisation, large or small, regardless of its field of activity. In fact, ISO 9001:2008 is implemented by over one million companies and organisations in over 170 countries.

Quality management principles

The standard is based on a number of quality management principles including a strong customer focus, the motivation and implication of top management, the process approach and continual improvement. Using ISO 9001:2008 helps ensure that customers get consistent, good quality products and services, which in turn brings many business benefits.

A quality management system is intended to help a company to:

- Identify, develop and implement efficient management systems
- Reduce waste to acceptable levels
- Ensure effective team working
- Regularly audit and review the Quality Management system to identify excellence, problems and areas for improvement

Promoting customer loyalty by the following:

- Ensuring that customers' needs are fully identified and understood
- Ensuring that project requirements are identified, understood, documented and agreed by all
- Providing a management system that ensures on time delivery of the agreed product or service
- Providing brief, user friendly and easily accessible methods to express satisfaction or dissatisfaction and ensure that any dissatisfaction is resolved
- Meeting the national statutory and regulatory requirements

To provide a good working environment and culture for staff by:

- Promoting a culture of honesty, timely communication and assistance to each other
- Providing brief, user friendly and easily accessible procedures and processes that reflect the users' preferred method of working wherever practicable
- Ensuring that recognition is given to deserving staff

Audits

Checking that the system works is a vital part of ISO 9001:2008. An organisation must perform internal audits to check how its quality management system is working.

An organisation may decide to invite an independent certification body to verify that it is in conformity to the standard, but there is no requirement for this. Alternatively, it might invite its Clients to audit the quality system for themselves.

Clause 40.2 requires the Consultant to provide, within the period stated in the Contract Data:

- a quality policy statement and
- a quality plan for acceptance.

Quality policy statement

A Consultant's quality policy statement, as required by Clause 40.2, defines its management's commitment to achieving and maintaining quality. It is often just a single paragraph, but it should clearly describe the Consultant's general quality orientation and clarify its basic intentions.

Quality policy statements should be used to generate quality objectives and should serve as a general framework for action. Quality policies can be based on the ISO 9000 Quality Management Principles and should be consistent with the organisation's other policies.

A typical quality policy statement could say:

> KH Consultants LLP is committed to achieving and consistently exceeding its Client's expectations, getting it right first time every time, by providing consultancy services to the highest quality and within a safe and environmentally friendly manner.
>
> Our success in achieving these objectives is continuously measured through our quality management system accredited under ISO 9001.

Quality plan

A quality plan, as required by Clause 40.2, is a document that is used to specify what procedures and processes will be followed in providing the specific service for the specific Client, who is responsible for managing them, to ensure that the quality expectations of the Client will be met as required by the Contract.

This may consist of the following sections:

- The Client's quality expectations – What the contract requires.
- Acceptance criteria – How quality will be measured.
- Quality responsibilities – Who is responsible.
- References to any quality standards – What other codes of practice, legislative and/or other standards apply, perhaps by law.
- Quality controls and audit processes – How the procedures and processes will be checked for compliance.
- Change management and procedures – How changes will be incorporated into the plan.

The plan should be then signed by all relevant parties to confirm that they have read and fully understood their responsibilities.

The degree to which the Service Manager checks conformity with the statement and plan is entirely a matter for the Service Manager, and perhaps the Client himself.

However, under Clause 40.3, if the Service Manager discovers a non-compliance, the Consultant is required to correct such failure on the instruction of the Service Manager.

Correcting Defects

It is important to stress that the Consultant is responsible for meeting the quality standards required by the contract and has an obligation to correct Defects whether or not the Service Manager has notified him of them.

In that respect, he should adopt a quality assurance/quality control system (as stated above), which prevents Defects occurring in the first place and corrects them promptly if they do occur.

Under Clause 41.1, until the defects date, the Service Manager is required to notify the Consultant of each Defect as soon as he finds it, and the Consultant is required to notify the Service Manager of each Defect as soon as he finds it (see Figure 4.1).

Contract: .... **Contract No: ....**	**DEFECTS NOTIFICATION** **Defects Notification No.................. .**
The Service Manager/Consultant is notified of the following defects:	
<u>Description</u>	**<u>Date Corrected</u>**
<u>The defect correction period is:</u>	
Certified all defects corrected................................(Service Manager)	**Date: ....**

Copied to:

Service Manager ☐ Consultant ☐ Other .... ☐ File ☐

Figure 4.1 Sample defects notification form.

The Consultant also notifies the Service Manager of the Defects he has corrected.

It is perhaps curious that the Consultant has to notify the Service Manager of each Defect when he finds it, when the Service Manager probably does not need to know about each Defect and if he did, he surely wants to know that the Consultant has corrected the Defect rather than just found it.

Note that the clause is referring to Defects in the services, not in the Works that have been carried out by a Contractor. If a Contractor has carried out defective work, then it should be fairly apparent what he has to do to correct the Defect i.e. replace something, rebuild it or carry out some form of remedial work to it to make good.

However, a Consultant having to correct a Defect raises a number of complex issues, for example, if there is a Defect in the design of something that the Consultant designed and the Contractor has now built, how can the Consultant correct the Defect and what is his liability in respect of possible demolition and rebuilding?

If the Consultant has been carrying out a survey and has missed an item from the survey, it is a Defect, but how can the Service Manager calculate the cost of having the Defect corrected by other people?

One assumes that the Service Manager calculates the amount in a fairly arbitrary manner, and the Consultant pays this as a form of damages, without anyone actually correcting it.

A Defect may occur for which the Consultant is not liable (see earlier note under definition of Defect). Nevertheless, the Consultant must correct the Defect, but it is a compensation event under Clause 60.1(13).

Defects Date

The Defects Date is set within Contract Data Part 1 and is normally fifty-two weeks after the completion of the whole of the service.

This is similar to the "Defects Liability Period", "Rectification Period" or "Defects Notification Period" in other contracts, the period within which the Consultant is initially liable for correcting Defects.

Defects Certificate

Unlike the Engineering and Construction Contract (ECC), there is no Defects Certificate under the Professional Service Contract; though again, many Clients include one within the contract so that any outstanding Defects are brought to the Consultant's attention.

Note that the Defects Certificate is not conclusive in that if a Defects Certificate states that there are no Defects, it does not prevent the Client from rights should a Defect arise later or if the Client did not find or notify it.

The Service Manager will normally only list what are usually defined as "Patent Defects" i.e., those which are observable from reasonable inspection at the time, and may not include what are usually defined as "Latent Defects", which may be hidden from reasonable inspection and may come to light at a later date, examples being some structural Defects.

The Consultant's and others' liability for correction of latent Defects and other costs associated with them will be dependent on the applicable law, and liability may remain despite the issue of any Defects Certificate.

5 Payment

Introduction

Payment, including assessment, certification and payment of amounts due, is covered within the NEC4 Professional Service Contract by Clauses 50.1–53.4, with further references within the Main Option Clauses.

There are separate payment provisions within the Professional Service Contract, dependent on the main option chosen (A, C or E), each option referring to two main terms:

i "The Prices"
These are the various elements that make up the total price for carrying out the service and will be in the form of an activity schedule, target or some other document dependent on the main option selected.

ii "The Price for Service Provided to Date"
This term is used in making the assessment of amounts due to the Consultant and is again dependent on the main option chosen.

Option A	Total of the Prices for each completed activity.
Option C	Total Defined Cost which the Service Manager forecasts will have been paid by the Consultant before the next assessment date plus the Fee.
Option E	Total Defined Cost which the Service Manager forecasts will have been paid by the Consultant before the next assessment date plus the Fee.

Payment under Options A, C and E

Option A – Priced contract with activity schedule.

The Prices are the lump sums for each of the activities on the Activity Schedule unless later changed in accordance with the contract.

The Price for Service Provided to Date (PSPD) is the total of the Prices for completed activities.

It is important to note that the Activity Schedule is the basis for the Consultant to be paid, it is not Scope or a schedule of what the Consultant has to do or is intending to do in Providing the Service.

Example

The Consultant has been awarded an Option A contract to design 50 No. houses, the Scope including submitting drawings and supporting information to the planning authority on behalf of the Client to obtain planning permission.

However, within the Activity Schedule the Consultant submitted with his tender, he has not specifically included an activity for submitting the information for planning permission.

The submission of the drawings for planning permission is deemed to be included within the Consultant's price and therefore must be assumed to have been included within the Activity Schedule. The same rule would apply to an Option C contract.

The Activity Schedule should also relate to operations on the Consultant's programme, and if, possibly, due to a change of method of working it does not, the Consultant must submit a revision to the Activity Schedule to the Service Manager for acceptance.

There will usually be more activities on the Activity Schedule than operations on the programme. This is because, for an operation which spans across a number of months, the Consultant will include a number of activities.

Example of an activity schedule

A Consultant has been appointed to carry out a property survey of one hundred existing houses for a local authority Client, the work being programmed to take four months.

The field survey will be carried out in months 1 and 2 and the compilation and submission of results being carried out in months 3 and 4.

That part of his Activity Schedule which covers the field survey could then look like this:

Complete field surveys

	£
Houses 1–10	8,500
Houses 11–20	8,850
Houses 21–30	8,500
Houses 31–40	7,950
Houses 41–50	8,750
Houses 51–60	7,950
Houses 61–70	8,500
Houses 71–80	8,850
Houses 81–90	7,950
Houses 91–100	8,850

Activities should be based on deliverables, so it is better to say "complete field surveys" rather than "carry out field surveys".

It is vitally important that when the Consultant submits the activity schedule, each activity is clearly defined to prevent debates between himself and the Service Manager about what an activity is and whether it is completed.

The Price for Service Provided to Date is then the total of the Prices for each *completed* activity in the Activity Schedule.

NB: The contract does not provide for incomplete activities to be paid.

A completed activity is one without Defects which would delay immediately following work.

So, in the example above, if the Consultant has completed the field survey to Houses 11–20, but there is a Defect within the survey that needs to be corrected, then as long as the Defect would not delay the following activities, then the activity is completed.

The Consultant is not entitled to be paid for activities which are not complete at the assessment date, it is "all or nothing", assessing the Consultant's right to payment under the contract is therefore different to the traditional approach of valuing the works carried out to date.

If the effect of a compensation event is to delay the completion of an activity, the Consultant is not entitled to be paid a proportion of the price for that activity, the delayed payment having to be priced within the quotation for the compensation event that caused the delay.

If a Consultant has establishment costs such as setting up site, providing accommodation, time-based preliminaries and similar activities as well as physical activities on site, it is important that the activities on the activity schedule are clearly defined not only in terms of what the work is, but where it is, so that their completion is not in doubt.

It is important on projects where the Consultant supplies preliminaries items, to correctly allocate these items to the Activity Schedule, as they could contribute as much as 10–15% of the Total of the Prices.

The traditional method of forecasting cash flow is by considering the programme and pricing on a month by month against the programme, inserting values against the events as they occur, allowing for delivery of materials as they are expected to arrive, and this is essentially the way that Option A works.

Option C – Target contract

As with Option A, the Prices are the lump sums for each of the activities on the Activity Schedule unless later changed in accordance with the contract.

The Price for Service Provided to Date is the total Defined Cost which the Service Manager forecasts will have been paid by the Consultant before the next assessment date plus the Fee.

This is quite a radical shift from the NEC3 Professional Services Contract where the payment was simply based on the "Time Charge", which was calculated from staff rates priced by the Consultant within Contract Data Part 2.

Note that within the Professional Service Contract, there is now provision under Options C and E for Disallowed Cost (see Disallowed Cost section below).

Many practitioners question the role of the Activity Schedule under Option C, when unlike Option A, it is not used for assessing payments. The answer is that the Activity Schedule under Option C assists the Client in assessing tenders as they can see how the Consultant has priced his tender.

Again, as with Option A, the Activity Schedule should relate to operations on the Consultant's programme, and if due to a change of method of working it does not, the Consultant must submit a revision to the Activity Schedule to the Project Manager for acceptance.

Option E – Cost-reimbursable contract

Option E was formerly known within the NEC3 Professional Services Contract as a "time based contract".

The Prices are the forecast of the total Defined Cost for the whole of the service plus the Fee.

The method of assessing payments is the same as Option C, i.e. the Price for Service Provided to Date is the total Defined Cost which the Service Manager forecasts will have been paid by the Consultant before the next assessment date plus the Fee, but there is no Consultant's share.

Disallowed Cost (Options C and E Only)

Disallowed Cost is cost which:

- is not justified by the Consultant's accounts and records

  The Consultant is obliged to keep accounts, proof of payments, communications regarding payments and any other records as stated in the Scope (Clauses 52.3 & 52.4).

  If it does not have the accounts and records to prove a cost, then the cost must be disallowed. It is not sufficient to merely prove that something physically exists for the Service Manager to see, as with many other contracts.
- should not have been paid to a Subcontractor or supplier in accordance with its contract

  The Consultant is obliged to submit the proposed subcontract documents for each Subcontractor to the Service Manager for acceptance, unless an NEC contract is proposed or the Service Manager has agreed that no submission is required.

  In addition, under Options C and E, the Consultant is required to submit the pricing information for each subcontract if an NEC contract is proposed, and the Service Manager instructs the Consultant to make the submission.

  If the Consultant pays a Subcontractor (or supplier) an amount that is not in accordance with its contract with them, then this is disallowed.
- was incurred only because the Consultant did not
  - follow an acceptance or procurement procedure stated in the Scope

    The Scope may require the Consultant to comply with a specific procedure in respect of, for example, submission of design proposals or procurement of Subcontractors. If the Consultant does not comply with the procedure, then associated costs are disallowed.
  - give an early warning which the contract required it to give

    If the Consultant incurs a cost that could have been avoided if the Consultant had given early warning, then it is disallowed.
  - give notification to the Service Manager of the preparation for and conduct of an adjudication or proceedings of a tribunal between the Consultant and a Subcontractor or supplier

And the cost of

- correcting Defects after Completion

  Correction of defects *after* completion is a Disallowed Cost. However, correction of defects *before* Completion is not a Disallowed Cost. This is often a contentious issue with Clients and Service Managers objecting to paying the Consultant for correcting its own defects.

  However, one must consider this clause logically. The Consultant has quite possibly priced the risk of having to correct defects which are its liability within its original tender, on an Option C contract, for example, this price will be included within the target.

  Option C is a cost-reimbursable contract until Completion, and the only way for the Consultant to be paid for this risk (should it materialise) is through the Defined Cost process as Option C is a cost-reimbursable contract until Completion.

  It is not a compensation event, so the target and/or the Completion Date are not changed. The Consultant is correcting the defect in its own time and potentially reducing the share it could make at Completion.
- correcting Defects caused by the Consultant not complying with a constraint on how it is to Provide the Service stated in the Scope

  A constraint may be stated in the Scope, such as a prescribed method of working, but the Consultant does not comply, and later, a Defect arises due to the Consultant's failure to comply with that stated constraint.

  If the Consultant does not comply with this constraint and a Defect occurs, then the Consultant's cost of correcting the Defect is disallowed.
- resources not used to Provide the Service (after allowing for reasonable availability and utilisation)

  Consultant's staff not working productively (or at all) but being charged in full, beyond what is considered "reasonable", is a Disallowed Cost.

  Note that, if the Consultant is using more resources than it had planned and priced for, or its resources are inefficient, that is not a Disallowed Cost.

  The question of what is "reasonable" can often be debatable. As with all Disallowed Cost, it is the Service Manager's responsibility to make the decision and to disallow the cost.
- preparation for and conduct of an adjudication, payments to a member of the Dispute Avoidance Board or proceedings of the tribunal between the Parties.

  If an adjudication, arbitration or legal proceedings occur, then each party bears its own costs. This bullet was not introduced until NEC3 was published, so whilst, then, Disallowed Cost provisions has only just been included within the NEC4 Professional Service Contract, prior to NEC3, under an Engineering and Construction Contract, for example, a Contractor could, in theory, refer a dispute to adjudication and whether or not it was successful, any costs arising could be included as Defined Cost and would not be disallowed.

Accounts and records

The Consultant is required to keep accounts and records of his Defined Cost (Options C and E) and, if applicable, expenses (Option A, C and E) at any time within working hours.

Under Option A, the Consultant is required to prepare forecasts of the total expenses for the whole of the services on a periodic basis and to submit them to the Service Manager (Clause 20.2) including an explanation of the changes since the last forecast.

Under Options C and E, the Consultant keeps accounts and records of his Defined Cost and expenses and allows the Service Manager to check them at any time during working hours (Clause 52.4).

In addition, the Consultant is required to prepare forecasts of the total Defined Cost and expenses for the whole of the service on a periodic basis and to submit them to the Service Manager (Clause 20.5) including an explanation of the changes since the last forecast.

The level of checking of the Consultant's accounts and records is at the Client's and Service Manager's discretion, some wish to examine each cost and its relevant back up document, some wish only to select certain costs at random, others wishing to only carry out a cursory check of costs.

Some Clients say that if the parties really are acting in a spirit of mutual trust and co-operation as the contract requires, then there should be no need for a detailed examination of the Consultant's accounts, though this probably needs a quantum leap of faith for many Clients.

It is important to establish the format on which accounts and records are to be presented by the Consultant to the Service Manager, particularly in respect of assessing payments, and this can be detailed in the Scope, and the following alternative methods are normally being employed:

i The Consultant submits a copy of all of his accounts and records to the Service Manager on a regular basis.

Or

ii The Consultant gives the Service Manager access to his accounts and records, and the Service Manager takes copies of all the records he requires. Whilst not expressly stated within the contract, the Service Manager should not have to travel too far to inspect them.

It has been known for accounts and records to be "available for inspection" in a different country to the site, the Consultant stating that they were available for inspection at any time during normal working hours. Again, the location and availability of accounts and records could be detailed within the Scope.

Or

iii The Consultant gives the Service Manager direct access to his computerised accounts and records through a PIN number, the Service Manager can then take copies of the records he requires.

The Consultant's share

The Client enters the Consultant's share percentage for each share range into Contract Data Part 1, the principle being that the Consultant receives a share of any saving and pays a share of any excess when the Price for Service Provided to Date is compared with the target at Completion of the whole of the works and in the final payment.

On completion of the service, the Service Manager makes a preliminary assessment of the Consultant's share by applying the Consultant's share percentage to the difference between the forecast final Price for Services Provided to Date and the forecast final total of the Prices. A final assessment is made using the final Price for Services Provided to Date and the final total of the Prices.

Whilst NEC practitioners often refer to the terms "pain" and "gain" in Option C, these terms do not actually exist within the contract. The Consultant pays or is paid the Consultant's share.

Example

At Completion of the service:

The Consultant's share percentages and the share ranges have been entered into Contract Data Part 1 as follows:

Share range	*Consultant's share percentage*
Less than 80%	50%
From 80–90%	40%
From 90–100%	30%
Greater than 100%	50%

The Total of the Prices	= £320,000
The Price for Services Provided to Date	= £265,500
Difference	= £54,500

The share range has been set in increments of 10%
10% of £320,000 = £32,000

"From 90% to 100%"
£32,000 × 30% = £9,600

"From 80% to 90%"
£22,500 × 40% = £9,000

Assessing the amount due

Under the NEC3 Professional Services Contract, the Consultant had to assess the amount which was due to him and submit an invoice to the Employer. The Employer could then notify the Consultant if he did not accept that assessment, then the Employer pays the amount due.

Under the NEC4 Professional Service Contract, the process is similar to the Engineering and Construction Contract in that the Service Manager (Project Manager under the Engineering and Construction Contract) assesses the amount due at each assessment date.

The first assessment date is decided by the Service Manager to suit the Parties and will normally be based on the time that the Consultant has been Providing the Service, the Client's procedures and timing for processing and issuing payments, and the Consultant's payment requirements and internal accounting system.

Clearly, in this respect, some discussion needs to take place between the Service Manager, the Client and the Consultant in order that a convenient assessment date can be set.

The first assessment must be made within the "assessment interval" after the starting date, and this is normally inserted in the Contract Data as "four weeks" or "one calendar month".

Later assessment dates occur at the end of each assessment interval until

- four weeks after the end of the defects date or
- the Service Manager issues a termination certificate.

Note that there is no provision within the contract for a minimum certificate amount.

The Service Manager assesses the amount due at each assessment date, calculating the Price for Service Provided to Date using the rules of the specific Main Option (A, C or E).

The Consultant submits an application for payment to the Service Manager before each assessment date and must include details of the application, the format to be in accordance with the Scope.

In making his assessment, the Service Manager considers "an application for payment the Consultant has submitted" before the assessment date.

If the Consultant submits an application for payment, the amount due to the Consultant is:

- the Price for Service Provided to Date (dependent on the Main Option used),
- plus, the amount of the expenses properly spent by the Consultant in providing the Service,
- plus, other amounts to be paid to the Consultant e.g. bonuses, Consultant's share, etc.,
- less amounts to be paid by or retained from the Consultant (e.g. retention, delay damages).

If the Consultant does not submit an application for payment, the amount due is the lesser of:

- the amount the Service Manager assesses as due at the assessment date, as though the Consultant *had* submitted an application for payment before the assessment date and
- the amount due at the previous assessment date.

In that case, if the Consultant does not submit an application for payment, the Service Manager's assessment would be ZERO or a MINUS figure i.e. an amount due to the Client.

It is essential that the Consultant either submits a programme with his tender or within the time scale specified within Contract Data Part 1. Failure to do so will entitle the Service Manager to retain one-quarter of the Price for Service Provided to Date in his assessment of the amount due.

This is a new provision not previously included within the NEC Professional Services Contracts.

Note that the amount is only withheld if the Consultant has not submitted a programme which shows the information which the contract requires e.g. method statement, time risk allowances, etc.

If the Consultant *has* submitted a programme which contains all the information that the contract requires, but the Service Manager disagrees with, for example, part of the method statement or the programme has not yet been accepted, then the provision, and the associated amount retained, does not apply.

Payment

The Service Manager is required to certify the payment due to the Consultant within one week of each assessment date (note the provisions under Clause Y(UK)2 if the contract falls under the Housing Grants, Construction and Regeneration Act 1996).

The payment (by either Party) is made by the later of

- one week after the paying Party (Client or Consultant) receives an invoice from the other Party (if an invoice is issued) and
- three weeks after the assessment date or, if a different period is stated in the Contract Data, within the period stated (see Figure 5.1).

Note, if the certificate is less than the previous one, then the Consultant pays the Client, the amount due, in which case the Client sends an invoice to the Consultant.

Payments are in the currency of the contract unless otherwise stated.

Interest is paid if a payment is not made or the Service Manager does not issue a certificate which he should have issued.

Similarly, interest is paid on a correcting amount due to corrections to a mistake or a compensation event or following a decision by the Adjudicator or the tribunal.

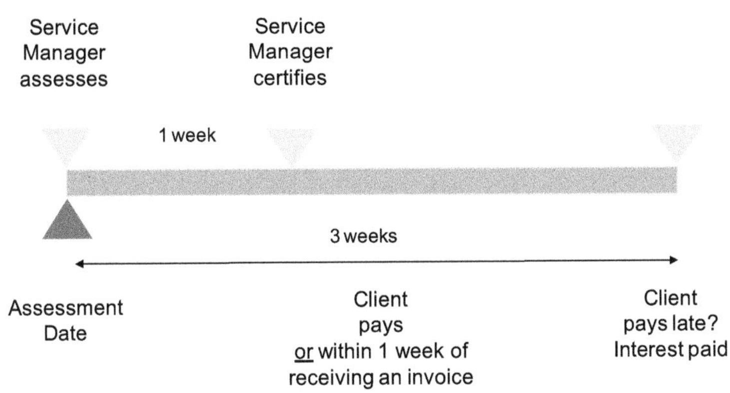

Figure 5.1 Time line for payments under the NEC4 Professional Service Contract.

The date from which interest should run is the date on which the additional payment would have been certified if there had been no mistake or dispute.

The interest rate is stated in Part 1 of the Contract Data and is assessed on a daily basis from the date the payment should have been made until the date when the late payment is made, calculated using the interest rate in Contract Data Part 1 compounded annually.

The interest due can be calculated on the basis of the following formula:

$$\frac{\text{Payment due} \times \text{interest rate} \times \text{the number of days late}}{365}$$

So, if one assumes the following:

- Payment due = £50,000
- Payment seven days late
- Interest rate in Contract Data Part 1 = 3%

The calculation is:

$$\frac{£50,000 \times 3\% \times 7 \text{ days}}{365}$$

Interest due = £28.77

Note that, if Option Y(UK)2 applies, if either Party intends to withhold payment of an amount due, he must notify the other not later than seven days before the final date for payment by stating the amount proposed to be withheld and the reason for withholding payment.

Also, if the Consultant exercises his right under the Housing Grants, Construction and Regeneration Act 1996 to suspend performance, it is a compensation event.

Schedule of Cost Components

Payment to the Consultant under Options C and E is now based on total Defined Cost which the Service Manager forecasts will have been paid by the Consultant before the next assessment date plus the Fee rather than the previous Time Charge.

Note that compensation events under Options C and E are also priced based on the Schedule of Cost Components.

There are two Schedules:

i Schedule of Cost Components
ii Short Schedule of Cost Components

The Schedule of Cost Components

Clause 1 – People

This relates to the cost of people who are directly employed by the Consultant and who is providing a part of the service.

The cost is calculated by taking the total cost and dividing by the total time recorded and then multiplying by the time spent in providing the service.

The cost component covers the full cost of employing the people including wages and salaries, contributions levies or taxes imposed by law, pensions and life insurance, death benefit, occupational accident benefits, medical aid and health insurance, and a vehicle.

It also includes bonuses and incentives, overtime and unsocial hours working, severance, protective clothing, safety training, relocation, medical examinations, passports and visas, travel insurance, including for dependants where appropriate.

Finally, it includes for the cost of people not directly employed by the Consultant but are paid according to time worked on the contract.

Clause 2 – Subcontractors

This component includes for payments to Subcontractors, without taking into account any amounts paid to or retained from the Subcontractor (such as retention) which would result in the Client paying or retaining the amount twice.

Clause 3 – Charges

This component includes a charge for support people (non-chargeable/non-fee earning) and office overhead costs and is calculated by applying the overhead percentage within the Contract Data to the total of the people costs in Item 1.

It includes for provision and use of people, accommodation, equipment, supplies and services required to provide the office and to support people providing the service.

Clause 4 – Insurance

The cost of events for which the Consultant is required to insure and other costs to be paid to the Consultant by insurers are *deducted* from cost. Note that insurance premiums (for example, Professional Indemnity (PI) Insurance) should be included within the percentage for Fee.

The Short Schedule of Cost Components

The Short Schedule of Cost Components is restricted to the assessment of compensation events under Option A, but if the Service Manager and Consultant agree, it may be used for assessing compensation events under Options C and E.

Clause 1 – People

This component is calculated by multiplying each of the People Rates by the total time appropriate to that rate properly spent on the contract.

Clause 2 – Subcontractors

This component covers payments to Subcontractors.

Clause 3 – Insurance

The cost of events for which the Consultant is required to insure and other costs to be paid to the Consultant by insurers are deducted from cost. Again, insurance premiums (for example, PI Insurance) should be included within the percentage for Fee.

The Fee

Whilst many practitioners believe that the Fee simply covers "overheads and profit", the definition is a little wider. All components of cost not listed in the Schedules of Cost Components are covered by the fee percentage.

There is no schedule of items covered by the fee percentage, but the following list, whilst not exhaustive, gives some examples of cost components not included in the Schedules of Cost Components.

i Profit
ii The cost of offices outside the Working Areas e.g. the Consultant's head office
iii Insurance premiums e.g. PI Insurance premiums
iv Performance bond costs
v Taxes
vi Advertising and recruitment costs
vii Sureties and guarantees required for the contract
viii Some indirect payments to staff

From this, one can see that there are some elements covered by the fee percentage which could be mistakenly assumed as being covered by the Schedules of Cost Components.

Final Assessment

The NEC Contracts have never had the equivalent of a Final Account or Final Certificate which is found in other contracts, certifying that the contract has fully and finally been complied with, and that issues such as payments, compensation events and defects have all been dealt with, and effectively, the contract can be closed.

Whilst some may say that because of the compensation event provisions, there is no Final Account, and then, there are provisions regarding finalisation of any outstanding Defects, and there is no need for a Final Certificate and there are some questions which could still remain such as when, under Option E, is it too late for the Consultant to submit a cost?

However, NEC4 has introduced a new provision where, in the case of the Professional Service Contract, the Service Manager makes a final assessment of amounts due to the Consultant, in effect giving closure, at least to financial aspects of the contract.

The Service Manager makes an assessment of the final amount due to the Consultant and certifies a final payment, no later than:

- four weeks after the later of the defects date, and the date the last defect is corrected, or
- thirteen weeks after the Service Manager issues a termination certificate.

The Service Manager gives the Consultant details of the assessment and payment, and the Party "to which payment is due" (which could be the Consultant or the Client) submits an invoice to the other Party (unusual to require an invoice?) with one week of that certificate, and the payment is made within one week of the other Party receiving the invoice or three weeks after the assessment date.

If the Service Manager does not make this final assessment within the time allowed, the Consultant may issue to the Client (not the Service Manager) his assessment of the final amount due.

If the Client agrees with the assessment, again, an invoice is issued by the Party to whom the payment is due and the payment is made within one week of the other Party receiving the invoice or three weeks after the assessment date.

The assessment of the final amount due is conclusive, unless:

If Option W1 is selected, a Party:

- refers a dispute about the assessment of the final amount due to the Senior Representatives within four weeks of the assessment being issued;
- refers any issues not agreed by the Senior Representatives to the Adjudicator within three weeks of the list of issues not agreed being produced, or when it should have been issued and
- refers to the tribunal its dissatisfaction with a decision of the Adjudicator regarding the final amount due within four weeks of the decision being made.

If Option W2 is selected, a Party:

- refers a dispute about the assessment of the final amount due to the Senior Representatives or to the Adjudicator within four weeks of the assessment being issued, but it may omit this stage by virtue of W2.2(1);
- refers any issues not agreed by the Senior Representatives to the Adjudicator within three weeks of the list of issues not agreed being produced, or when it should have been issued and
- refers to the tribunal its dissatisfaction with a decision of the Adjudicator regarding the final amount due within four weeks of the decision being made.

The assessment of the final amount due is changed to include any agreement the Parties reach and a decision of the Adjudicator or recommendation of the Dispute Avoidance Board which has not been referred to the tribunal within four weeks of the decision or recommendation.

A changed assessment becomes conclusive evidence of the final amount due.

6 Compensation events

Introduction

Compensation events, including their definition, notification, quotations, assessment and implementation, are covered within the NEC4 Professional Service Contract by Clauses 60.1–66.3 with further references within the Main Option Clauses.

The "traditional approach" to managing change

The "traditional (non-NEC) approach" to managing change in construction contracts is normally dealt with in three separate elements within the contract.

Variation

The changes in the form of additions, omissions or substitutions are priced by measuring the changed work, then pricing the change using rates consistent with the Consultant's original tender, the pricing document for that tender normally being in the form of lump sums or schedules of rates.

Where the changed work is similar to that referred to in the pricing document, executed under similar conditions, and does not significantly change the quantity of work set out in the tender, the original rates and prices apply. Where the changed work is similar to that referred to in the tender, but not executed under similar conditions, and/or there is a significant change in the quantity of work, the original rates and prices form the basis of the pricing i.e. they are adjusted to allow for the differences in conditions and/or quantities. There are normally provisions for the parties to agree fair rates and prices and where the work cannot be clearly defined and/or measured, then day work provisions apply.

Whilst this has always been seen as the "normal" way to price change in a construction contract, it has its drawbacks.

The first problem with the traditional approach is that the rates and prices inserted by the Consultant in his tender were calculated based on the original quantities, and whether they are right or wrong, they are being applied to changes which are outside his control. In that sense, either the Consultant or the Client may gain or lose by the changes.

The second problem is that there are rarely specific time scales within the contracts for submitting the prices and agreeing them, apart from the Final Account which is normally submitted by the Consultant and agreed by the Contract Administrator, months or even years after the works have been completed.

Extension of time

Whilst the variation element deals with the changes to the prices, the effect on the completion date under non-NEC contracts is dealt with as a *separate* process referred to as "extension of time" or "adjustment to completion date". This provision allows for the Contract Administrator acting for the Client to extend time/fix a new completion date, preventing time becoming at large and thereby postponing the Client's right to recover liquidated damages.

In that respect, the Consultant is required to submit a separate notice to the Contract Administrator upon it becoming apparent that a delay will occur. Normally, there is a time scale within these provisions for the Consultant to submit his notice and in turn for the Contract Administrator to respond.

The purpose of the notice requirement is partly to allow the Contract Administrator to attempt to mitigate the problem either by stopping the delay from recurring or by even omitting work in addition to making a judgement on the Consultant's entitlement to additional time.

The notice will normally identify:

a The cause of the delay – detail of how and why the delay is occurring or likely to occur.
b The estimated effect on completion – the Consultant should also give an estimate of delay in his notice so that the Contract Administrator can form his own opinion.
c The clauses on which the Consultant relies in requesting an extension of time.

iii Loss and Expense.

Finally, contracts have provision for the Consultant to recover loss and expense which he could not have recovered under any other provision within the contract.

Loss and expense claims have from experience been given a very bad name. In reality, loss and expense claims should be a process by which the Consultant can submit a well-structured, fully substantiated claim to seek to recover any loss or expense which he has incurred as a result of something which is outside his control and/or foreseeability.

Whilst the financial effect of a variation should essentially be priced with the variation, the cumulative effect of a number of variations or other factors which are not the Consultant's risk may involve the Consultant in loss and expense in having to work uneconomically, out of sequence, etc., and this provision deals with that possibility. Often, agreement to loss and expense claims is a prolonged process which extends long after completion of the project. They are generally known as disruption claims.

The NEC approach

The NEC contracts take a very different approach to pricing and managing change as contained within the compensation event provisions, whereby the event is notified, priced, assessed and implemented within a defined, relatively short time scale, including the cost and time effect of the change (see Figure 6.1)

NEC recognises that it is *critical* that the time and cost effect of a change are considered *together* as the result of that change; therefore, the Prices and the Completion

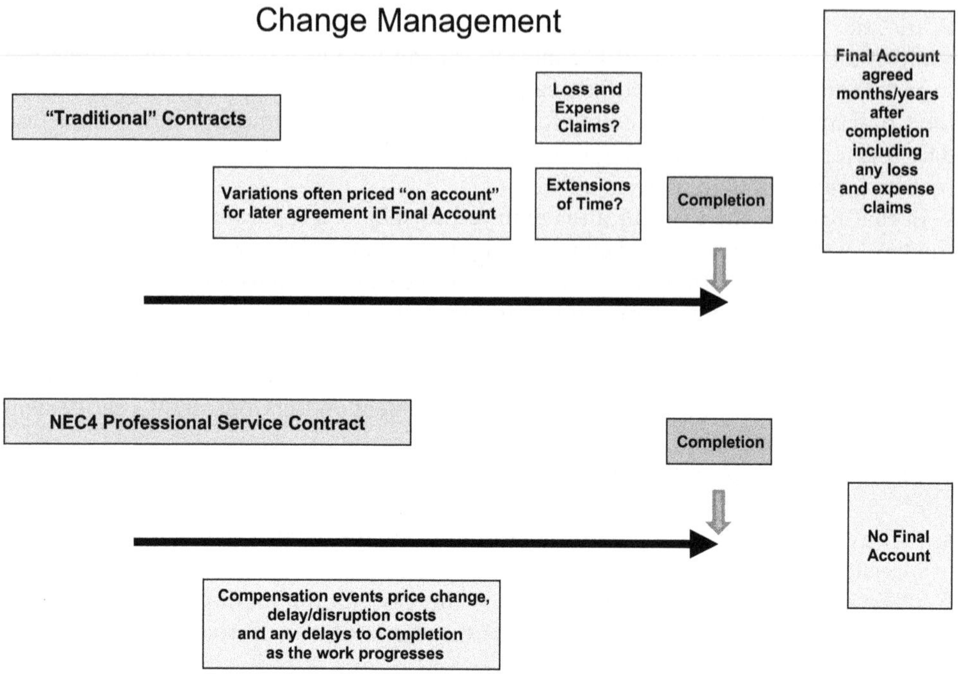

Figure 6.1 Change management under the NEC4 Professional Service Contract.

Date are changed within a single provision rather than within three separate processes as with the traditional approach above.

Unfortunately, experience shows that generally, the requirements of the compensation events procedures are poorly understood and applied by all parties.

The more onerous requirements on the Service Manager particularly in responding to quotations were introduced in NEC3 to meet a particular failing as were the time bar requirements on both Parties.

Definition of a compensation event

There is no definition of a compensation event within the contract; however, it could be defined as "an event which is the Client's risk, and if it occurs, and affects the Consultant, entitles the Consultant to be compensated for any effect on the Prices, the Completion Date, and any Key Date".

The compensation events

The compensation events may be found in a number of locations within the contract:

- Sixteen events are stated in sub-clauses 60.1(1)–60.1(16) – these are core clauses and therefore always apply, unless amended by Z clause provisions.

<table>
<tr><td>Contract: ....

Contract No: ....</td><td>SERVICE MANAGER'S INSTRUCTION

SMI No.......................................</td></tr>
<tr><td>To: Consultant</td><td>☐ Instruction changing Scope
☐ Proposal to change Scope
☐ Other</td></tr>
<tr><td colspan="2">Description

</td></tr>
<tr><td>Further instructions:
☐ This instruction is/is not a compensation event
☐ Provide quotation of time/cost effect
☐ Start work immediately
☐ Keep records</td><td>
☐ Issued as proposed change. Do not implement.
☐ Provide quotation of time/cost effect
☐ No cost instruction</td></tr>
<tr><td>Signed: (Service Manager) ....</td><td>Date: ....</td></tr>
<tr><td colspan="2">Consultant's Summary

☐ Delay to Completion is...............days/weeks
☐ No delay to Completion
☐ Change in Price Plus/Minus £ ....
☐ No change in Price
Note: This is a summary only. Full supporting documentation or records must accompany this statement.</td></tr>
<tr><td>Signed: (Consultant) ....</td><td>Date:...............</td></tr>
</table>

Copied to:

Service Manager ☐ Consultant ☐ Other .... ☐ File ☐

Figure 6.2 Sample service manager's instruction.

- Further events are included in secondary options X2 (Changes in the Law), X12 (Multiparty Collaboration) and Y(UK)2 (Housing Grants, Construction and Regeneration Act 1996).
- Additional compensation events may also be stated by included in Contract Data Part 1.
- Also, compensation events can be added, omitted or amended through Z clauses, though the preference is that they are included within Contract Data Part 1.

The compensation events within the core clauses are set down in Clause 60.1:

> 60.1(1) The Service Manager gives an instruction changing the Scope.

Under Clause 14.3, the Service Manager may give an instruction to the Consultant which changes the Scope or a Key Date (see Figure 6.2).

Changes to the Scope are normally defined in other contracts as "Variations". The NEC contracts do not use the term Variations, but the Service Manager can give an instruction changing the Scope, which can be in the form of an addition, omission or alteration to the service.

An instruction can also be given in order to resolve an ambiguity or an inconsistency within the Scope. An ambiguity may be defined as "an uncertain meaning or intention" and therefore is something that is unclear; an inconsistency may be defined as a "conflict in meaning or intention" and therefore could be a mismatch or discrepancy between documents.

Under Clause 16.1, the Service Manager or the Consultant notifies the other as soon as either becomes aware of the ambiguity or inconsistency, the Service Manager then gives an instruction resolving it and the subsequent change to the Scope is then a compensation event under Clause 60.1(1).

A compensation event which arises from an instruction to change the Scope to resolve an ambiguity or inconsistency is assessed "as if the Prices, the Completion Date and the Key Dates were for the interpretation most favourable to the Party which did not provide the Scope" (Clause 63.10), usually the Client, so the "benefit of the doubt goes in favour of the Consultant, who is deemed to have priced the cheaper, easier or quicker alternative".

Example

A part of the Scope shows that a traffic survey must be carried out twice a year, whilst another part of the Scope shows that the same survey must be carried out three times a year.

Probably, an early warning would be issued first under Clause 15.1. to seek clarification.

Under Clause 16.1, the Service Manager or the Consultant notifies the other as soon as either becomes aware of the ambiguity or inconsistency and the Service Manager then gives an instruction resolving it; the subsequent change to the Scope is then a compensation event under Clause 60.1(1) and is assessed under Clause 63.10 (above) as if the Prices, the Completion Date and the Key Dates were for the interpretation most favourable to the Party which did not provide the Scope, in this case, the Consultant.

If the Service Manager confirms that the Consultant must carry out the survey twice a year, then there would probably be no change to the Prices; if three times a year, the outcome would be that the Prices are changed to include the additional payment to the Consultant for the third survey, and possibly, there may be a delay to planned Completion.

Also, under Clause 16.2, the Consultant notifies the Service Manager as soon as he considers that the Scope requires him to do anything which is illegal or impossible, the Parties cannot contract to do something illegal. If the Service Manager agrees, he gives an instruction to change the Scope appropriately.

> 60.1(2) The Client does not allow access to a person, place or thing by the later of its access date and the date for access shown on the Accepted Programme.

The Client has a fundamental obligation to allow the Consultant access to "a person, place or thing as stated in the contract".

The term "thing" in a contract seems a somewhat curious word, but it could describe a location, article, facility or service to be provided by the Client.

The Service Manager should be aware that if the Consultant shows a date in his programme when he requires access to a person, place or thing, and the Service Manager accepts the programme, then the Service Manager is obliged to allow that access by the later of that date and the date in the Contract Data (Clause 33.1), failing which a compensation event occurs.

So, the Consultant can *relax* the obligation to the date in the contract by showing a later date in the Accepted Programme, but cannot impose an *earlier* date than that stated in the Contract.

Note that the NEC3 Professional Services Contract only referred to access "as stated in this contract" rather than by reference to the Accepted Programme.

> 60.1(3) The Client does not provide something which it is to provide by the date shown on the Accepted Programme.

The Client, through the Service Manager, is obliged to provide information, facilities or other things as shown on the Accepted Programme.

It is important that in addition to the Service Manager stating the dates when things will be provided, the Consultant shows these requirements and the dates for providing them in his programme submitted for acceptance so that when the Service Manager accepts the programme, he accepts liability on behalf of the Client if he fails to provide it.

Note that, whilst the sub-clause refers to "the Client", it applies to anyone who acts for the Client and has an obligation to provide something, so this clause could include, for example, the Service Manager not providing information to the Consultant, or another consultant employed by the Client not providing design information to the Consultant.

> 60.1(4) The Service Manager gives an instruction to stop or not to start any work, or to change a Key Date.

It is important to note that the Service Manager giving an instruction to stop or not to start any work or to change a Key Date is always a compensation event, regardless of the reason why the instruction was given; however, if the Service Manager gives an instruction to the Consultant to stop a part of the service, for example, because the Consultant is working unsafely and the Consultant notifies the Service Manager that he believes it is a compensation event, then under Clause 61.4, if the Service Manager decides that the event arose from a fault of the Consultant, then he notifies the Consultant that the Price, Completion Date and the Key Dates are not to be changed and gives his reasons within the notification.

> 60.1(5) The Client or Others do not work within the times shown on the Accepted Programme or the conditions stated in the Scope.

The words "do not work within the times shown on the Accepted Programme" can often cause debate. What work is the Client, or Others, required to do? Also, "work within the times" infers a process over a period of time rather than a specific task or date, so it may be difficult to determine whether someone has worked within the times shown on the Accepted Programme or within the conditions stated in the Scope.

Again, it is important that in addition to the Client showing what he is going to do, the Consultant shows these requirements and the dates for providing them in his programme submitted for acceptance so that when the Service Manager accepts the programme, he accepts liability on behalf of the Client if he fails to provide it.

Note that the term "Others" are people who are not the Client, Service Manager, Consultant, a Subconsultant or the Adjudicator, examples of Others being contractors or other consultants, third-party regulatory authorities, utilities companies installing or diverting electricity, gas or water services and is a traditional area of difficulty.

Some Clients amend this clause to exclude their liability for "Others", thus passing the risk to the Consultant. This is unreasonable because the "Others" may be employed by the Client, or if a genuine third party, the Consultant has no more contractual or financial control over the service providers than the Service Manager, it is unfair that the Consultant should hold the risk.

> 60.1(6) The Service Manager does not reply to a communication from the Consultant within the period required by the contract.

If the Service Manager does not reply to a communication, which could be for example, the Consultant notifying an early warning, or some other notifictaion which it is required to give under the Contract and not receiving a reply, then it is a compensation event as it is fundamentally a breach of contract.

However, if the Consultant notifies a compensation event in this respect, and the Service Manager decides that event has no effect upon the Consultant's costs, Completion or meeting a Key Date, then again under Clause 61.4, he notifies the Consultant that the Price, Completion Date and the Key Dates are not to be changed.

The "period required by this contract" may either be an expressly stated period of time, for example, two weeks for the Service Manager to reply to the submission of programme (Clause 31.3) or the more general "period for reply" stated in Contract Data Part 1.

60.1(7) The Service Manager changes a decision which the Service Manager had previously communicated to the Consultant.

The contract refers to "decisions" under the following clauses, underlining added for emphasis:

- Clause 30.2: The Service Manager <u>decides</u> the date of Completion.
- Clause 61.5: The Service Manager <u>decides</u> that the Consultant did not give an early warning of the event which an experienced Consultant could have given.
- Clause 61.6: The Service Manager <u>decides</u> that the effects of a compensation event are too uncertain to be forecast reasonably.
- Clause 64.1: The Service Manager assesses a compensation event if he *decides* that the Consultant has not assessed the compensation event correctly.

This clause has been interpreted by some NEC practitioners as a provision to allow the Service Manager to change and revise his decision regarding the assessment of an implemented compensation event.

This is an incorrect interpretation; it must be stressed that the Service Manager does not have the authority to change his assessment once it has been implemented. The Service Manager's assessment is final, and the only recourse would be adjudication.

60.1(8) The Service Manager withholds an acceptance (other than acceptance of a quotation for acceleration or of not correcting a Defect) for a reason not stated in the contract.

The clauses dealing with acceptance are as follows:

- Clause 21.2: Acceptance of a proposed key person or a replacement
- Clause 23.2: Acceptance of a proposed Subcontractor
- Clause 23.3: Acceptance of the proposed subcontract documents for a Subcontractor
- Clause 31.3: Acceptance of a programme
- Clause 32.2: Acceptance of a revised programme
- Clause 35.3: Acceptance of a quotation for an acceleration
- Clause 62.3: Acceptance of a Consultant's quotation for a compensation event (Option A)
- Clause 55.3: Acceptance of a revised activity schedule (Options C and E)
- Clause 24.4: Acceptance of the pricing information for each subcontract (Secondary Options)
- X4: Acceptance of an ultimate holding company guarantee
- X10: Acceptances regarding Information Modelling
- X13: Acceptance of a performance bond
- X20: Acceptances regarding Key Performance Indicators

Within each of the clauses are stated the reasons for not accepting. If the Service Manager does not accept for a different reason, then this is a compensation event.

For example, the Service Manager may not accept a proposed Subcontractor as he has never heard of them. The Consultant may offer additional information about this proposed Subcontractor, but if the Service Manager still does not accept them as he has never heard of them, then the reason for not accepting is not that the appointment of the Subconsultant "will not allow the Consultant to Provide the Service".

> 60.1(9) An event which is a Client's liability stated in these conditions of contract.

The Client's liabilities under Clause 80.1 (Liabilities and Insurance) are:

- claims and proceedings from Others and compensation and costs payable to Others which are due to
  - the unavoidable result of the service and
  - negligence, breach of statutory duty or interference with any legal right by the Client or by any person employed by or contracted to it except the Consultant, e.g. the Service Manager or other consultants;
- a fault of the Client or any person employed by or contracted to it, except the Consultant and
- additional Client's liabilities stated in the Contract Data.

Any loss or damage will be covered by insurance, but any delay to Completion or effect on the Prices cost will be recovered as a compensation event.

> 60.1(10) The Service Manager notifies the Consultant of a correction to an assumption which the Service Manager stated about a compensation event.

Under Clause 61.6, if the Service Manager decides that the effects of a compensation event are too uncertain to be forecast reasonably, he states assumptions, the Consultant's quotation being based on these assumptions (see example under Clause 61.6).

The Service Manager's assumption may state what conditions the Consultant should assume so the Consultant can price it, for example, a delay and/or the effect on Defined Cost (plus Fee) which the Consultant is to assume for the time being. Sometimes, the Service Manager will tell the Consultant exactly what Defined Cost to allow.

If the Service Manager states what Defined Cost (plus Fee) the Consultant is to allow, practitioners often refer to this as the Service Manager giving the Consultant a "Provisional" Sum for the time being, and in effect that is what he is doing, though one should not use the term "Provisional Sum" as the NEC contracts have never provided for them, for very good reasons.

If the Service Manager later notifies a correction to the assumption, it is a compensation event under Clause 60.1(10). In this case, the change to the Prices and any delay to Completion the Consultant had assumed in his original quotation are then replaced with corrected amounts.

Example

The Service Manager states that the Consultant should assume the Defined Cost of £3,000 for a part of a quotation for a compensation event which is too uncertain to be forecast reasonably.

The actual Defined Cost of this element is subsequently recorded by the Consultant and found to be £3,400.

The Service Manager notifies a correction to the assumption under Clause 60.1(10), which is then priced as follows:

	£
Defined Cost from records	3,400
Service Manager's Assumption	3,000
	400
Plus, Fee say 10%	40
Change to the Prices:	+440

60.1(11) A breach of contract by the Client which is not one of the other compensation events in this contract.

This is a "catch all" to cover any breaches not covered by the other compensation events, an example could be the Service Manager has not paid a certificate and the effect of the non-payment is that the Consultant has to delay or stop progress.

60.1(12) An event which

- stops the Consultant completing the whole of the service or
- stops the Consultant completing the service by the date shown on the Accepted Programme

and which

- neither Party could prevent
- an experienced consultant would have judged at the Contract Date to have such a small chance of occurring that it would have been unreasonable for him to have allowed for it and
- is not one of the other compensation events stated in this contract

Clause 60.1(12) is a "no fault" clause introduced under NEC3, and dealing with an event neither Party could prevent, which stops the Consultant completing the services. This clause is also tied in with Clause 18.1 "Prevention"

18.1 If an event which

- stops the Consultant completing the whole of the service or
- stops the Consultant completing the whole of the service by the date for planned Completion shown on the Accepted Programme

and which

- neither Party could prevent and
- an experienced consultant would have judged at the Contract Date to have such a small chance of occurring that it would have been unreasonable to have allowed for it:

The Service Manager gives an instruction to the Consultant stating how he is to deal with the event.

In that sense, it requires the Service Manager to come up with the solution and give the appropriate instruction.

There has been much confusion and debate regarding these clauses since NEC3 was first published in 2005.

The intention of the drafters appears to be that this would be a force majeure (superior force) clause, often referred to as an "Act of God" in other contracts and legal documents i.e. significant events such as flooding, earthquakes, volcanic eruptions and the associated dust clouds and other natural disasters which prevent the contracting parties from fulfilling their obligations under the contract, the clause essentially freeing them, or giving some relief, from their liabilities.

Whilst this appears to be what the drafters had intended, on closer examination, the application of this clause is far wider:

If you take each element in turn:

The event

- stops the Consultant completing the whole of the service.
  Due to the event, the service was never completed.
- stops the Consultant completing the whole of the service by the date for planned Completion shown on the Accepted Programme.
  Due to the event, completion of the service was delayed.
- neither Party could prevent
  Clearly, if either Party *could* have prevented the event, then they *would* or *should* have. The clause refers to *preventing* the event rather than taking some mitigating action to lessen its effect.
- an experienced consultant would have judged at the Contract Date to have such a small chance of occurring that it would have been unreasonable to have allowed for it.
  At tender stage, the likelihood of the event occurring was so low, the Consultant would not have allowed for it, but how would you define and measure a "small chance"?
- is not one of the other compensation events stated in this contract?
  If the matter could have been dealt with as one of the other compensation events, then it should be. Clause 60.1(12) then clearly deals with the exceptional event.

As examples of where this clause would at least cause confusion:

The Professional Service Contract has no equivalent of the Engineering and Construction Contract Clause 60.1(13) related to weather conditions, so one would then assume that the risk of unforeseen weather conditions is the Consultant's and he should have allowed for it.

But Clause 60.1(11) could, at least in theory, change this?

The Consultant's progress with an external survey is affected by a very severe storm which occurs less frequently than once in 50 years, which delays completion of the works by 1 week.

The Consultant states as follows:

- The weather stopped him completing the survey by the date shown on the Accepted Programme. Assume that the Consultant can prove that.
- Neither party could prevent the event. Whilst one can take measures to lessen the impact of a storm, one cannot prevent the storm from occurring.
- An experienced Consultant would have judged at the Contract Date to have such a small chance of occurring that it would have been unreasonable to have allowed for it. The Consultant states that he could not have predicted or allowed for such a rare event and its consequences.
- It is not one of the other compensation events. There is no weather-related compensation event in the contract, but has Clause 60.1(11) now created one?

Could Clause 60.1(12) also extend to loss or damage occasioned by insurable risks such as fire, lightning, explosion, storm, flood, earthquake and terrorism?

If Clause 60.1(12) is intended as a form of force majeure provision rather than including a fairly long-winded definition within the clause, why not just state:

"A force majeure event occurs" and allow the parties to interpret this internationally recognised term as they do in other contracts.

> 60.1(13) The Consultant corrects a Defect for which he is not liable under the contract.

The Consultant is obliged to correct Defects; however, a Defect may occur for which the Consultant is *not* liable (see earlier note under definition of Defect). If the Consultant corrects a Defect for which he is not liable, it is a compensation event.

> 60.1(14) The Service Manager gives an instruction correcting information provided by the Client.

This is reasonably straightforward in that any information provided by the Client, which is incorrect, would have to be corrected, and this would be a compensation event.

> 60.1(15) The Service Manager notifies the Consultant that a quotation for a proposed instruction is not accepted.

This new compensation event is related to Clause 65.3, where the Consultant has submitted a quotation for a proposed instruction, and the Project Manager has replied to the quotation stating that the quotation is not accepted and that the Project Manager may issue an instruction, notify the instruction as a compensation event and instruct the Consultant to submit a quotation for additional work that may not go ahead.

It gives the Consultant some recompense for preparing and submitting quotation for the proposed instruction.

> 60.1(16) Additional compensation events stated in Contract Data part one.

This new compensation event enables any additional compensation events to be detailed within the Contract Data rather than as Z clauses.

Weather-related risks (see below) could be detailed within the Contract Data.

As stated above, there are further compensation events within the following Secondary Options:

- Option X2 Changes in the law – i.e. when a change in law occurs after the Contract Date.
- Option X12 Multiparty collaboration – changes to the Partnering Information and changes to the plan caused by amendments to the timetable.
- Option Y(UK)2 The Housing Grants, Construction and Regeneration Act 1996 – Consultant suspends performance due to late/non-payment.

Comparison of compensation events under the NEC3 Professional Services contract and the NEC4 Professional Service Contract

There are sixteen compensation events within the core clauses of the NEC4 Professional Service Contract and twelve within the core clauses of the NEC3 Professional Services Contract (see Chart 6.1 at the end of the chapter).

Barring the change from "Employer" to "Client" and the introduction of the "Service Manager" the differences are the following four compensation events which exist under the NEC4 Professional Service Contract and *not* under the NEC3 Professional Services Contract:

60.1(9) An event which is a Client's liability stated in these conditions of contract

60.1(14) The Service Manager gives an instruction correcting information provided by the Client

60.1(15) The Service Manager notifies the Consultant that a quotation for a proposed instruction is not accepted

60.1(16) Additional compensation events stated in Contract Data part one

Comparison of compensation events under the NEC4 Professional Service Contract and the NEC4 Engineering and Construction Contract

There are sixteen compensation events within the core clauses of the NEC4 Professional Service Contract and twenty-one within the core clauses of the NEC4 Engineering and Construction Contract (see Chart 6.2 at the end of the chapter), plus others to be found within the Main and Secondary Options.

Fourteen of the compensation events within the NEC4 Professional Service Contract (Clauses 60.1(1)–60.1(12), 60.1(15) and 60.1(16)) are virtually the same as the NEC4 Engineering and Construction Contract, the only differences being the reference to different parties.

There are then a further seven compensation events which are within the NEC4 Engineering and Construction Contract and *not* the NEC4 Professional Service Contract which relate to a construction contract and one which is in the NEC4 Professional Service Contract but not in the NEC4 Engineering and Construction Contract which relate to a consultancy agreement.

Weather-related risks

As can be seen from the comparison above, unlike the NEC4 Engineering and Construction Contract, the NEC4 Professional Service Contract does not include a compensation event for weather-related risks.

Whilst this may at first not seem to be an important issue for the Consultant, for example, in a drawing office designing a project rather than the Contractor on site building it, it can be a concern and a major risk for a Consultant carrying out other professional services which may be affected by the weather, for example, site inspections, surveys, etc.

Example

On an Option A contract, a Consultant is employed to carry out an inspection of existing drainage systems within a region, which is estimated will take him three weeks. However, in the second week, heavy rain impedes progress and also causes flooding which means that the inspection is delayed, and the inspection is eventually completed in four weeks (i.e. one-week late).

The Consultant will be liable for his own costs in any delays and possibly demobilising/remobilising due to the bad weather, but if Option X7 is selected, he would also be liable to the Client for payment of delay damages.

Where external work is involved, it may be appropriate for the Client to add a Z clause for weather-related risks adapted from Clause 60.1(13) of the NEC4 Engineering and Construction Contract i.e.

> 60.1(13) A weather measurement is recorded
>
> - within a calendar month
> - before the Completion Date for the whole of the services and
> - at the place stated in the Contract Data
>
> the value of which, by comparison with the weather data, is shown to occur on average less frequently than once in ten years.
>
> Only the difference between the weather measurement and the weather which the weather data show to occur on average less frequently than once in ten years is taken into account in assessing a compensation event.

However, as discussed above, one should consider whether Clause 60.1(12) would allow a compensation event for the weather, or any other event, which is outside the Consultant's control, foreseeability, etc. and is not one of the other compensation events under the Contract.

Notifying compensation events

Compensation events may be notified to the Consultant by the Service Manager or to the Service Manager by the Consultant (Clause 61) (see Figure 6.3).

It is important that the correct procedures for notifying and assessing compensation events are followed.

Contract: Contract No:	COMPENSATION EVENT NOTIFICATION CE No..
To: Service Manager/Consultant	
You are notified of the following compensation event: Description The Compensation Event is likely to have an effect on: ☐ Price ☐ Completion Has an early warning been given? ☐ Yes ☐ No Quotation Required? ☐ Yes ☐ No	
Signed: Service Manager/Consultant):..............................	Date: ....

Copied to:

Service Manager ☐ Consultant ☐ Other .... ☐ File ☐

Figure 6.3 Sample compensation event notification.

There are two ways of notifying compensation events (note the NEC3 Professional Services Contract including a proposed instruction, but that has now been made a separate provision (Clause 65)):

- If a compensation event arises from the Service Manager giving an instruction, or notification or changing an earlier decision, the Service Manager notifies the Consultant of the compensation event at the time of the communication (Clause 61.1). He also instructs the Consultant to submit quotations, unless the event arises from a fault of the Consultant or quotations have already been submitted.

It is important to note that if the Consultant has been given an instruction by the Service Manager to do something, he should comply with that instruction and not wait until the quotation has been submitted or accepted, so the Consultant puts the instruction or changed decision into effect. Whether that instruction constitutes a compensation event should also not affect the carrying out of that instruction.

- The Consultant can notify a compensation event, but must do so within eight weeks of becoming aware of the event; otherwise, he is not entitled to a change in the Prices, the Completion Date or a Key Date, unless the Service Manager should have notified the event to the Consultant but did not (Clause 61.3).

The Consultant must then respond within one week or whatever period the Consultant agrees, as to whether it is a compensation event.

The intention of the "eight week rule" is to compel the Consultant to notify compensation events promptly; otherwise, any entitlement to additional time and money is lost.

Note that the clause requires the Consultant to notify the compensation event to the Service Manager within eight weeks of becoming aware of it, not just to have given early warning or to mention the possibility of a compensation event in a discussion with the Service Manager.

The clause therefore covers compensation events initiated by the Consultant rather than the Service Manager, which the Service Manager may not have been aware of unless the Consultant had notified it.

An example of the Service Manager failing to notify a compensation event when he should have would be if an instruction was issued by the Service Manager to the Consultant changing the Scope, but at the time the Service Manager did not notify the compensation event, and the Consultant in turn did not notify it either. Even if the compensation event is not notified within eight weeks, the responsibility remains with the Service Manager as he gave the instruction and should have notified the event to the Consultant but did not.

Various opinions have been published about the enforceability and effectiveness of time bars in contracts such as NEC, commentators particularly debating whether the clause is a condition precedent to the Consultant being able to recover time and money, and whether a party, the Service Manager, can benefit from its own breach of contract to the detriment of the injured party, the Consultant.

For example, if the Service Manager does not provide something which he is to provide by the date for providing it shown on the Accepted Programme, this is a valid compensation event under Clause 60.1(3), but can the Service Manager prevent the Consultant from receiving any remedy because the Consultant failed to notify the Service Manager within the eight-week limit, and possibly, if Completion is delayed, the Consultant have to pay delay damages.

The author, whilst not being a lawyer, is of the opinion that the parties are clear as to what the terms of their agreement are at the time the contract is formed. It is also clear what happens if the Consultant does not notify a compensation event within the time stated within Clause 61.3; he is protected against notifications which the Service Manager should have given but did not, and therefore the time bar must be effective and enforceable.

Clause 61.4 covers the possible outcomes to the Consultant's notification of a compensation event under Clause 61.3.

Negative reply from the Service Manager

If the Service Manager responds by stating that an event notified by the Consultant

- arises from a fault of the Consultant
- has not happened and is not expected to happen
- has not been notified within the time scales set out within the Contract
- has no effect upon the Defined Cost, Completion or meeting a Key Date or
- is not one of the compensation events stated in the contract

He notifies the Consultant of his decision that the Prices, Completion Date and the Key Dates are not to be changed. The Service Manager needs to name one of these as his reason for refusing a compensation event not to simply say that he does not believe that it is a compensation event.

Positive reply from the Service Manager

If the Service Manager decides otherwise, he notifies the Consultant that it is a compensation event and instructs him to submit quotations.

No reply from the Service Manager

There has always been concern amongst Consultants who were working under the first and second editions of the Professional Service Contract, what remedy was available if the Service Manager (previously the Employer) failed to give *any* reply to a notification of a compensation event within the period required by the contract, the simple answer being that failure to reply to a communication within the period required by the contract is another compensation event under Clause 60.1(6), which does not remedy the problem with the initial notification of a compensation event.

The drafters of NEC3 considered the problem and introduced an amendment new within Clause 61.4. If the Service Manager does not reply to a notification of a compensation event within one week, the Consultant may notify him to that effect.

If the Service Manager does not reply within two weeks of the notification, the notification from the Consultant is treated as acceptance that the event is a compensation event and an instruction to submit quotations.

Remedy for failure to give early warning

While there is no requirement for an early warning to be issued for a matter which has already been notified of a compensation event, there is a sanction where the Consultant does not give warning of something that may later become a compensation event.

Under Clauses 61.5 and 63.7, if the Service Manager decides that the Consultant did not give an early warning of the event which an "experienced" consultant could have given, he notifies this to the Consultant when he instructs him to submit quotations. If the Service Manager has done so, the event is assessed as if the Consultant had given early warning (see example in Chapter 1).

What this means in practice is that it is not in the interests of the Consultant to hold back from giving an early warning of something that may later become a compensation event.

Service manager's assumptions

When the Consultant is instructed to submit a quotation, there may be a part of the quotation which is too uncertain to be forecast reasonably. In this case, the Service Manager should state what the Consultant should assume, which could be a financial and/or time effect.

As stated previously, the Service Manager's assumption may state what conditions the Consultant should assume or could even state a Defined Cost which the Consultant is to assume for the time being.

Subsequently, when the effects are known or it is possible to forecast reasonably, he notifies a correction to the assumption, and it is dealt with as a correction under Clause 60.1(10).

Note, the Service Manager must state the assumption under Clause 61.6, if the Consultant makes assumptions within his quotation when pricing the compensation event, then they are not corrected later.

Another example of a Service Manager's assumption.

Example

The Consultant has been appointed to carry out site investigations for a proposed new development and has been instructed by the Service Manager to excavate a trial pit to establish the nature of the sub strata in a part of the Site, so that the design for a new foundation for an oil storage tank can be finalised. The Service Manager tells the Consultant that he and the structural engineer will inspect the sides to the excavation as the Consultant progresses with the work and will instruct the Consultant when to stop excavating.

As the plan size and depth of the excavation is uncertain, the Consultant is unable to make a reasonable forecast of proposed changes to the Prices and any delay to Completion, so the Service Manager states that the Consultant should allow an excavation

4 m × 4 m and to a depth of 3 m and to backfill the excavation on completion when instructed. The Consultant submits his quotation which is accepted by the Service Manager, the compensation event is then implemented.

Subsequently, the excavation is actually required to be 4 m × 4 m and to a depth of 4.5 m.

The Service Manager therefore notifies a correction to an assumption, which is a compensation event under Clause 60.1(9) and instructs the Consultant to submit a quotation. The resulting quotation deducts the Defined Cost assumed for the original trial pit and forecasts the Defined Cost for the actual trial pit, and this is the change to the Prices, with any delay to Completion also adjusted.

Quotations for compensation events

Once the Service Manager has instructed the Consultant to submit a quotation, the Consultant has three weeks in which to submit his quotation to the Service Manager (note under the NEC3 Professional Services Contract, the Consultant had two weeks to submit his quotation to the Service Manager (see Figure 6.4), although the period may be extended by agreement between the Service Manager and the Consultant under Clause 62.5., but it has to be agreed before the Consultant's submission or the Service Manager's reply is due.

If the Service Manager and Consultant do not agree to extend the time, once the three weeks have expired, the Service Manager notifies the Consultant that he will be

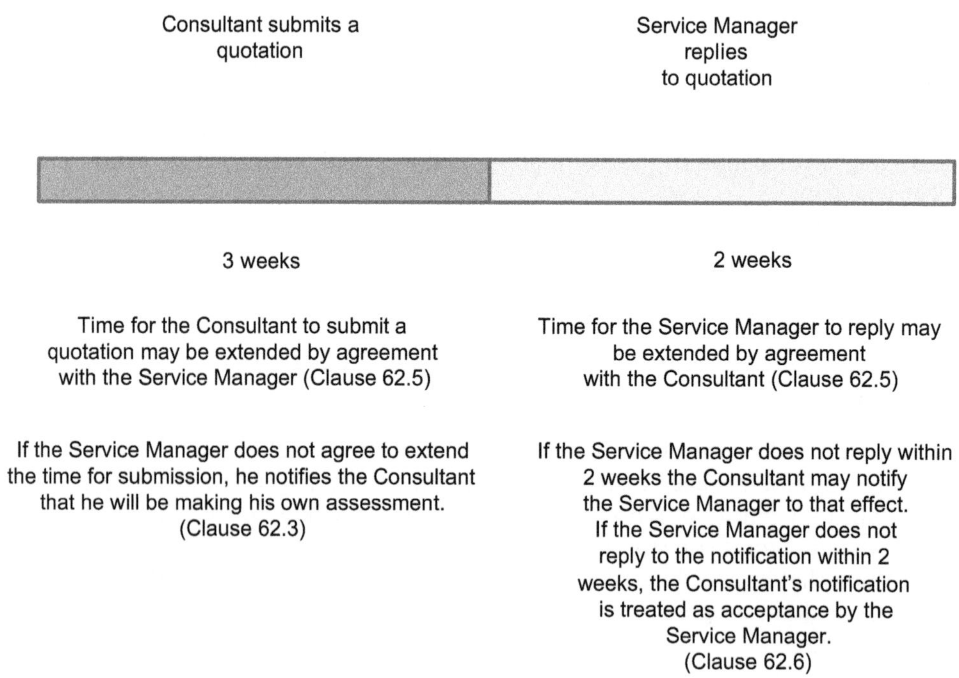

Figure 6.4 Times for submission and assessment of quotations for compensation events.

making his own assessment (Clause 62.3). If the Consultant does not submit a quotation, the Service Manager notifies within two weeks of the date the Consultant should have submitted it.

Once the Consultant has submitted his quotation to the Service Manager, the Service Manager is required to reply as follows:

- he could accept the quotation; in which case, it is implemented (Clause 66.1)
- he could instruct the Consultant to submit a revised quotation, in which case, he explains his reasons for doing so to the Consultant. The Consultant then has a further three weeks to submit a revised quotation (Clause 62.4), though in most cases, by agreement, he will review and re-submit within a shorter time scale as both Parties need to know the outcome
- he could notify the Consultant that he will be making his own assessment, in which case, the Service Manager has up to three weeks to make that assessment from the time of the notification

Many practitioners believe that the Service Manager has to actually make the assessment within the initial two-week period for response. This is incorrect; the Service Manager only has to notify the Consultant within the two-week period that he will be making his own assessment, and the details of the assessment must be notified to the Consultant; within the time period, the Consultant had to submit the quotation i.e. normally within a further three weeks (Clause 64.3).

If the Service Manager makes an assessment and the Consultant disagrees with that assessment, then the Consultant can refer the matter to Adjudication.

NB There is no contractual provision in any of the NEC contracts to "negotiate and agree" the Prices.

Normally, quotations consist of both money and time implications (Clause 62.2).

The Consultant is also free to offer other alternatives such as using different methods.

Delays to completion

Note that under Clause 62.2, within the Consultant's quotation, he needs to consider and include any delay to the Completion Date. Note that the clause states "any delay" and not "any effect", so the Completion Date can only be changed to a later date or remain unchanged due to a compensation event.

From this provision, one can see that no compensation event can result in an earlier Completion Date, though it may give rise to an earlier planned Completion.

If the Service Manager requires an earlier Completion Date, due, for example, to an omission of work, then he should consider the acceleration provisions within the contract (Clause 35).

This could also be used in an effort to *maintain* a Completion Date, despite the Consultant being required to carry out additional work which for a compensation event.

Note that in assessing any delay to the Completion Date, the delay is the length of time that due to the compensation event planned Completion is later than planned Completion as shown on the Accepted Programme (Clause 63.5), so any terminal float attached to the whole programme is retained by the Consultant.

Similarly, any delay to a Key Date is the length of time that due to the compensation event planned date when the Condition will be met is later than the planned date as shown on the Accepted Programme.

The assessment takes into account any delay caused by the compensation event already in the Accepted Programme and events that have happened between the date of the Accepted Programme and the dividing date.

Clearly, this emphasises the need for an Accepted Programme to be in place in order that it can be used to make correct assessments.

Any time, risk allowances which the Consultant has allowed are preserved by this clause, as assessment of the compensation event is based on "entitlement" rather than "need". Allowances for risk must be included in forecasts of Defined Cost and Completion in the same way that the Consultant allows for risks when pricing his tender. Float within the Accepted Programme is, however, available to mitigate or avoid any consequential delay to the Completion Date.

It is worth noting that the Consultant will usually show as little float as possible in his Accepted Programme, i.e. all operations are close to or on the critical path so that any compensation event delays Completion and in turn the Completion Date. This has been highlighted in Chapter 3 as one of the reasons; it is important for the Service Manager to spend time gaining a full understanding of the Consultant's programme and properly evaluating it before accepting it.

Alternative quotations

Under Clause 62.1, the Service Manager may require the Consultant to provide alternative quotations, for example, to carry out the work using existing resources or, alternatively, provide additional resources which might mitigate delay. There could be the alternative of a quick but expensive solution versus a slower but cheaper solution.

Discussion will normally take place between the Service Manager and the Consultant to ascertain what is feasible and practicable.

No reply to a quotation

As stated above, there has always been concern amongst Consultants who were working under the first and second editions of the Professional Service Contract, what remedy was available if the Service Manager failed to reply to a notification of a compensation event, and also to the submission of a quotation within the period required by the contract, the simple answer being that failure to reply to a communication within the period required by the contract is another compensation event under Clause 60.1(6), which, again does not remedy the problem with the initial compensation event.

The drafters of NEC3 considered the problem and introduced a new Clause 62.6, which, in seeking to provide a remedy to the Consultant for the Service Manager's failure, seems to favour the Service Manager.

If the Service Manager does not reply to a quotation within the time allowed, the Consultant may notify him to that effect. If the Consultant has submitted more than one quotation, he states in that notification which one he proposes is to be accepted.

If the Service Manager does not reply within two weeks of the notification, the notification from the Consultant is treated as acceptance of the Consultant's quotation by the Service Manager (Clause 62.6).

It is somewhat concerning that a pre-condition to deemed acceptance is that the Consultant must issue a notification giving the Service Manager a further, but final two weeks in which to deliberate, yet if the Consultant fails to provide a quotation within the time allowed, the Service Manager's notification tells the Consultant that he will be making his own assessment, so there is no further time for the Consultant to respond.

In addition, Secondary Option W1 (Dispute Resolution) specifically allows the Service Manager to refer a dispute about a quotation for a compensation event which is treated as having been accepted, to be referred to the Adjudicator, so the Consultant still does not have final acceptance! This is a curious inclusion when Clause 62.6 states that no reply to the Consultants notification is treated as acceptance by the Service Manager. One would assume that an Adjudicator whose role is to enforce the contract would have to find in favour of the Consultant.

Assessing compensation events

The basic rule for assessing compensation events is stated in Clause 63.1.

This is based on the effect of the event upon:

- the actual Defined Cost of the work done by the dividing date,
- the forecast Defined Cost of the work not done by the dividing date and
- the resulting Fee.

Note that the NEC3 Professional Services Contract used the term Time Charge rather than Defined Cost (plus Fee) so in addition to changing Time Charge to Defined Costs which is now the same as the Engineering and Construction Contract, the new term "dividing date" was included within the NEC4 Professional Service Contract.

Assessment of compensation events is based on their effect on Defined Cost and time, Defined Cost as defined within the Schedule of Cost Components or the Short Schedule of Cost Components.

This is different from most standard forms of contract where variations are valued using the rates and prices in the contract as a basis. The reason for this policy within the Engineering and Construction Contract is that no compensation event which is the subject of a quotation is due to the fault of the Consultant or relates to a matter that is at its risk under the contract. It is therefore appropriate to reimburse the Consultant its forecast additional costs or Defined additional costs arising from the compensation event.

It is also worth some clarification here regarding the term "dividing date".

Clause 63.1 in previous editions of the NEC3 Professional Services Contract stated:

- the actual Time Charge of the work already done
- the forecast Time Charge of the work not yet done and
- the resulting Fee

This led to some confusion as to when, in preparing his quotation, the Consultant changed from "actual Time Charge" to "forecast Time Charge", so in addition to changing Time Charge to Defined Costs, which is then the same as the Engineering and Construction Contract, the new term "dividing date" was included within the NEC4 Professional Service Contract.

Again, as previously stated above, for a compensation event arising from the Project Manager or Supervisor giving an instruction or notification, issuing a certificate or changing an earlier decision, the "dividing date" is the date of that communication.

For other compensation events, the "dividing date" is the date of the notification of the compensation event.

So, in preparing his quotation, the Consultant uses actual Defined Cost of work that has been done by the dividing date and forecast Defined Cost of the work yet to be done by the dividing date.

Note that if the effect of a compensation event is to reduce the Prices, the Prices are not reduced unless it is a change to the Scope (Clause 60.1) or a correction to an assumption (Clause 60.1(10)).

The assessment should also include for any forecast delay to the Completion Date.

Assessments should always allow for cost and time including allowances for risk (Clause 63.8) which has a significant chance of occurring and are the Consultant's risk under the contract.

This should include for anything which would not be classified as a separate compensation event e.g. the possibility of adverse weather conditions when carrying out external work.

Clearly, a properly constructed programme showing float and time risk allowances will assist in this exercise.

Example

In the Option A contract, the Service Manager gives an instruction to the Consultant as follows:

> Carry out structural inspections to an additional 10 No. properties as shown on Drawing No. 32.

The Consultant's quotation will comprise changes to the Prices and also any delay to the Completion Date and any Key Dates in the contract.

Delay to completion

First, the Consultant considers whether there will be any delay to planned Completion. The Consultant should provide details to demonstrate to the Service Manager the basis on which the delay has been calculated, and this will probably require him to issue a programme or an extract from the programme with his quotation (Clause 62.2).

He assesses that this additional work will take him two weeks to complete, and as the instruction has been issued only one week before planned Completion, he assesses that it will delay Completion by one week.

Delay to completion has then been assessed as one week.

Changes to the prices

The changes to the Prices are not made by using the Consultant's rates from the activity schedule. They must be assessed by forecasting the Defined Cost for the additional work already done and/or to be done and adding the Fee.

The Consultant must also ensure that he includes for any "establishment/preliminaries costs" for the one week delay to planned Completion.

The Consultant assesses that this additional work will require an additional building surveyor for two weeks, plus two days of a structural engineer's time to analyse the results. There will also be some nominal print costs.

In addition, the hire of the temporary office accommodation they have been using for the duration of the contract will have to be extended by one week (see Figures 6.5 and 6.6).

COMPENSATION EVENT QUOTATION EXAMPLE	
Option C Contract	
Carry out structural inspections to an additional 10 No. properties as shown	
on Drawing No. 32	
The Consultant assesses that this will require an additional Building Surveyor	
for two weeks, plus 2 days of a Structural Engineer's (Subcontractor) time to	
analyse the results.	
In addition, the Consultant needs to factor in the cost of equipment and to enable	
him to carry out the surveys.	
Under Option C, Defined Cost is calculated by reference to Schedule of Cost	
Components:	
People	
NB this should be to cost to employ staff, not a "charge out rate" which may	
include profit	
Building Surveyor (directly employed) - 1 x 55 hours x £50.00 per hour	2750.00
Charges	
Cost of support people, overhead costs including accommodation, equipment,	
supplies and services, etc.	
People cost x overhead percentage from Contract Data Part 1 (assume 8%)	
£2750.00 x 8%	220.00
Subcontractor	
Structural Engineer (subcontracted) - 1 x 16 hours x £90.00 per hour	540.00
	3510.00
Add Fee percentage (assume 10%)	351.00
Total of Quotation:	**3861.00**

Figure 6.5 "Sample Compensation Event Quotation" form.

Contract: Contract No:	Compensation Event Quotation CE No......................................
To: Service Manager	
<u>Summary of Quotation</u>	
Change to Prices:	
Change to Completion Date:	
Signed: .... (Consultant)	Date: ....
The Service Manager's reply is: ☐ An Instruction to submit a revised quotation ☐ Acceptance of the quotation ☐ A notification that a proposed instruction will not be given ☐ A notification that the Service Manager will be making his own assessment	
Signed: .... (Service Manager)	Date: ....

Copied to:

Service Manager ☐ Consultant ☐ Other .... ☐ File ☐

Figure 6.6 "Sample Compensation Event Quotation Submission" form.

Changes to the scope that omit services

When the Service Manager gives an instruction changing the Scope by omitting work, the method of changing the Prices and dealing with the Completion Date is often misunderstood.

The tendency is for parties to simply delete or even to just ignore the item in the pricing document, for example, in an Option A contract, the activity in the activity schedule which represents the omitted work is simply reduced or not paid.

Omitted work due to a change to the Scope is assessed not by omitting or remeasuring the item in the relevant pricing document, but by forecasting the Defined Cost of that omitted work. The resulting amount is then adjusted against the value in the pricing document.

This principle is clearly stated in Clause 63.1 of the contract, i.e. the changes to the Prices are based on:

- the actual Defined Cost of the work done by the dividing date,
- the forecast Defined Cost of the work not done by the dividing date and
- the resulting Fee.

It must be remembered that the principle with compensation events in terms of their financial value is that *neither* party gains or loses as a result of the compensation event; the Consultant is compensated so that he is in the same financial position *after* the event as he would have been *before* the event.

If a Consultant has included low rates in his tender for work which is subsequently omitted it is quite probable that the application of forecast Time Charge will give rise to a negative value; this is testament to the fact that the Consultant would have made a loss if the work had *not* been omitted. Similarly, if high rates exist, the Consultant will retain the margin he would have made if he had carried out the work.

Example

In an Option A contract to design a new office building, the Activity Schedule includes an activity entitled "design reception area", the activity having a price of £32,000. The Consultant at the time of tender was unsure what the design would involve so assumed the value of £32,000.

The Service Manager gives an instruction to omit the design of the reception area as he has decided to employ a specialist interior designer under a separate contract to carry out this part of the work; this is a change to the Scope and therefore a compensation event under Clause 60.1(1).

The tendency is to assume that the Consultant is simply not paid for that activity as it will not be carried out; however, the correct way to assess the change to the Prices is to forecast the Time Charge of the work not yet done and this becomes "the change to the Prices".

Let us assume that the Consultant assesses the Defined Cost plus Fee for the omitted work at £28,500.00, then that is "the change to the Prices".

The amount of the quotation, assuming that it has been accepted by the Service Manager, is then used to change the Prices, so either a new activity is then inserted to the value of "−£28,500.00" or the activity priced at £32,000.00 is omitted and replaced with an activity priced at £32,000.00 − £28,500.00 = £3,500.00.

The Consultant in this case retains the profit he would have made if the design to the reception area had not been omitted. Clearly, if the change to the Prices had been more than £32,000, the Consultant would retain the loss.

The Service Manager's assessments

Under Clauses 64.1–64.4, the Service Manager may, for any of the following reasons, make his own assessment of a compensation event:

- if the Consultant has not submitted a required quotation and details of his assessment within the time allowed,
- if the Service Manager decides that the Consultant has not assessed the compensation event correctly in a quotation and he does not instruct the Consultant to submit a revised quotation,
- if, when the Consultant submits quotations for a compensation event, the Consultant has not submitted a programme which the contract requires him to submit or
- if, when the Consultant submits quotations for a compensation event, the Service Manager has not accepted the Consultant's latest programme for one of the reasons stated in the Contract.

These are all derived from some failure of the Consultant either to submit a quotation, to assess the compensation event correctly or to submit an acceptable programme.

If the Service Manager makes his own assessment, he should put himself in the position of the Consultant, giving a properly reasoned assessment of the effect of the compensation event, detailing the basis of his calculations and providing the Consultant with details of that assessment. A Service Manager's assessment is not simply the quotation returned to the Consultant with "red pen" reductions down to a figure the Service Manager is prepared to accept or can afford to pay.

It is also important to recognise that the Service Manager is not sending his assessment to the Consultant for his acceptance, and it is the final decision under the contract, the Consultant's only remedy being adjudication.

Under Clause 64.4, if the Service Manager does not assess a compensation event within the time allowed, the Consultant may notify him to that effect. If the Service Manager does not reply to the notification within two weeks, the Consultant's notification is treated as acceptance of the quotation by the Service Manager.

The only remedy the Service Manager has, if he later finds that the quotation is not acceptable is to refer it to adjudication. However, it must be remembered that the role of the Adjudicator and the adjudication process is to enforce the contract; therefore, if the Consultant's quotation has been submitted in accordance with the contract, then the Service Manager will probably be unsuccessful in the adjudication.

Proposed instructions

Under Clause 65, the Service Manager may instruct the Consultant to submit a quotation for a proposed instruction. Note that the Consultant has only been instructed to submit a quotation, and he does not carry out the work until instructed to do so.

The Consultant submits his quotation within three weeks of being instructed to do so.

Once the Consultant has submitted his quotation to the Service Manager, the Service Manager is required to reply as follows:

- he could instruct the Consultant to submit a revised quotation; in which case, he explains his reasons for doing so to the Consultant. The Consultant then has a further three weeks to re-submit it, though, in most cases, he will review and re-submit within a shorter time scale;
- he could issue an instruction to carry out the work and accept the quotation; in which case, it is implemented;
- he could notify the Consultant that the quotation is not accepted.

If the quotation is not accepted, the Service Manager may still instruct the work to be carried out, but it is treated as a compensation event and the Consultant is instructed to submit a quotation.

Implementing compensation events

Under Clause 66, the Service Manager implements each compensation event by notifying the Consultant of the quotation which he has accepted or of his own assessment, or a quotation is treated as having been accepted.

The Parties should keep a continual summary of all compensation events which have occurred throughout the contract and the progress of each one (see Figure 6.7).

This is normally kept electronically and circulated to all relevant parties.

The assessment of a compensation event is not revised if a forecast on which it was based is shown by later information to have been wrong, and this clause emphasises the finality of the assessment of compensation events.

- Under Option A, assessments for changed Prices are in the form of changes to the Activity Schedule. The Consultant is then paid that amount when the activity is completed.
- Under Option C, again assessments for changed Prices are in the form of changes to the Activity Schedule.

However, as Option C is a target contract, the assessment only adjusts the target (Total of the Prices); the Consultant is then actually paid the Time Charge for work which has been completed, with the Consultant's share calculated at Completion.

- Under Option E, the assessment is only used for forecasting purposes and does not affect the payments made to the Consultant

Final certificate

It is worth mentioning that the NEC Contracts have never had the equivalent of a "Final Account" or "Final Certificate" which is found in other contracts, certifying

Contract: .... Contract No:				SUMMARY OF COMPENSATION EVENTS				
Ref.	Date notified	Origin SM or Cons	Description of event	Clauses	Ref.	Action by SM or Cons	Date Implemented	Notes
1								
2								
3								
4								
5								
6								
7								
8								
9								
10								
11								
12								
13								

Service Manager ☐ Consultant ☐ Other .... ☐ File ☐

Figure 6.7 Sample record of compensation events.

that the contract has fully and finally been complied with, and that issues such as payments and compensation events have all been dealt with.

There is now a provision for a Final Assessment, which is discussed in detail within Chapter 5.

Appendix

Suggested templates for notifying and managing compensation events (see Figures 6.6 and 6.7 and Charts 6.1 and 6.2).

NEC3 Professional Services Contract (PSC)

60.1(1)	*The Employer gives an instruction changing the Scope.*
60.1(2)	*The Employer does not provide access to a person, place or thing for the Consultant as stated in this contract*
60.1(3)	*The Employer does not provide something which he is to provide by the date for providing it shown on the Accepted Programme.*
60.1(4)	*The Employer gives an instruction to stop or not to start any work or to change a Key Date.*
60.1(5)	*The Employer or Others do not work within the times shown on the Accepted Programme or within the conditions stated in the Scope.*
60.1(6)	*The Employer does not reply to a communication from the Consultant within the period required by this contract.*
60.1(7)	*The Employer changes a decision which he has previously communicated to the Consultant.*
60.1(8)	*The Employer withholds an acceptance (other than acceptance of a quotation for acceleration) for a reason not stated in the contract.*
60.1(9)	*The Employer notifies a correction to an assumption which he has stated about a compensation event.*
60.1(10)	*A breach of contract by the Employer which is not one of the other compensation events in this contract.*
60.1(11)	*An event which stops the Consultant completing the services..........*
60.1(12)	*The Consultant corrects a Defect for which he is not liable under this contract.*

Chart 6.1 Comparison of compensation events under the NEC3 Professional Services Contract and the NEC4 Professional Service Contract.

NEC4 Professional Service Contract (PSC)

60.1(1) The Service Manager gives an instruction changing the Scope.

60.1(2) The Client does not allow access to a person, place or thing by the later of its access date and the date for access shown on the Accepted Programme

60.1(3) The Client does not provide something which he is to provide by the date shown on the Accepted Programme.

60.1(4) The Service Manager gives an instruction to stop or not to start any work, or to change a Key Date.

60.1(5) The Client or Others do not work within the times shown on the Accepted Programme or the conditions stated in the Scope.

60.1(6) The Service Manager does not reply to a communication from the Consultant within the period required by the contract.

60.1(7) The Service Manager changes a decision which the Service Manager has previously communicated to the Consultant.

60.1(8) The Service Manager withholds an acceptance (other than acceptance of a quotation for acceleration or of not correcting a Defect) for a reason not stated in the contract.

60.1(9) An event which is a Client's liability stated in these conditions of contract stated about a compensation event.

60.1(10) The Service Manager notifies the Consultant of a correction to an assumption which the Service Manager stated about a compensation event.

60.1(11) A breach of contract by the Client which is not one of the other compensation events in this contract.

60.1(12) An event which stops the Consultant completing the services……….

60.1(13) The Consultant corrects a Defect for which he is not liable under the contract.

60.1(14) The Service Manager gives an instruction correcting information provided by the Client.

60.1(15) The Service Manager notifies the Consultant that a quotation for a proposed instruction is not accepted.

60.1(16) Additional compensation events stated in Contract Data part one.

Chart 6.1 (Continued).

NEC4 Professional Services Contract (PSC)

60.1(1) The Service Manager gives an instruction changing the Scope.

60.1(2) The Client does not allow access to a person, place or thing by the later of its access date and the date for access shown on the Accepted Programme

60.1(3) The Client does not provide something which he is to provide by the date shown on the Accepted Programme.

60.1(4) The Service Manager gives an instruction to stop or not to start any work, or to change a Key Date.

60.1(5) The Client or Others do not work within the times shown on the Accepted Programme or the conditions stated in the Scope

60.1(6) The Service Manager does not reply to a communication from the Consultant within the period required by the contract.

60.1(7) The Service Manager changes a decision which the Service Manager has previously communicated to the Consultant.

60.1(8) The Service Manager withholds an acceptance

60.1(9) An event which is a Client's liability stated in these conditions of contract stated about a compensation event.

60.1(10) The Service Manager notifies the Consultant of a correction to an assumption which the Service Manager stated about a compensation event.

60.1(11) A breach of contract by the Client which is not one of the other compensation events in this contract.

60.1(12) An event which stops the Consultant completing the services……….

60.1(13) The Consultant corrects a Defect for which he is not liable under the contract.

60.1(14) The Service Manager gives an instruction correcting information provided by the Client.

60.1(15) The Service Manager notifies the Consultant that a quotation for a proposed instruction is not accepted.

60.1(16) Additional compensation events stated in Contract Data part one.

NEC4 Engineering and Construction Contract (ECC)

60.1(1) The Project Manager gives an instruction changing the Scope except

60.1(2) The Client does not allow access to and use of a part of the Site by the later of its access date and the date for access shown on the Accepted Programme.

60.1(3) The Client does not provide something which it is to provide by the date for providing it shown in the Accepted Programme.

60.1(4) The Project Manager gives an instruction to stop or not to start any work or to change a Key Date.

Chart 6.2 Comparison of compensation events under the NEC4 Professional Service Contract and the NEC4 Engineering and Construction Contract.

60.1(5) The Client or Others do not work within the times shown on the Accepted Programme, etc

60.1(6) The Project Manager or the Supervisor does not reply to a communication from the Contractor within the period required by the contract.

60.1(7) The Project Manager gives an instruction for dealing with an object of value or of historical or other interest.

60.1(8) The Project Manager or the Supervisor changes a decision which has previously communicated to the Contractor.

60.1(9) The Project Manager withholds an acceptance

60.1(10) The Supervisor instructs the Contractor to search for a Defect

60.1(11) A test or inspection done by the Supervisor causes unnecessary delay

60.1(12) The Contractor encounters physical conditions

60.1(13) A weather measurement is recorded

60.1(14) An event which is a Client's liability stated in these conditions of contract

60.1(15) The Project Manager certifies take over of a part of the works before both Completion and the Completion Date.

60.1(16) The Client does not provide materials, facilities and samples for tests and inspections as stated in the Scope.

60.1(17) The Project Manager notifies a correction to an assumption which the Project Manager stated about a compensation event.

60.1(18) A breach of contract by the Client which is not one of the other compensation events in this contract.

60.1(19) An event which stops the Contractor completing the whole of the works

60.1(20) The Project Manager notifies the Contractor that a quotation for a proposed instruction is not accepted.

60.1(21) Additional compensation events stated in Contract Data Part 1.

Compensation Events that exist within the NEC4 Engineering and Construction Contract, but not the NEC4 Professional Service Contract.

60.1(7) *The Project Manager gives an instruction for dealing with an object of value*

60.1(10) The Supervisor instructs the Contractor to search for a Defect and no Defect is found.........

60.1(11) A test or inspection done by the Supervisor causes unnecessary delay.

60.1(12) The Contractor encounters physical conditions...............

60.1(13) A weather measurement is recorded...........

60.1(15) The Project Manager certifies take over of a part of the Works before both Completion and the Completion Date.

60.1(16) The Employer does not provide materials, facilities and samples for tests and inspections as stated in the Scope.

Chart 6.2 (Continued).

7 Rights to material

Introduction

Rights to material are covered within the NEC4 Professional Service Contract by Clauses 70.1–70.3.

The parties' use of material

The Consultant retains the rights over any material which he produces, but the Client has effectively a licence to use it for the purposes stated in the Scope (Clause 70.1). Similarly, the Consultant obtains equivalent rights from the Subconsultant to his material.

However, if Secondary Option X9 (Transfer of rights) is included in the contract, these rights are transferred to the Client with exceptions being stated in the Scope; so, effectively, the position is reversed. Again, the Consultant obtains equivalent rights for the Client from any Subcontractor to his material. The Consultant is also required to provide documentary evidence of this transfer.

Confidentiality is achieved by the provisions of Clause 70.2.

Many Clients also include specific Z clauses related to title in respect of:

- the Official Secrets Act
- non-disclosure agreements (NDA)
- requirements for security passes
- prohibition of all photographs
- patents and copyrighted materials
- business practices
- client lists
- computer data

The Consultant is free to use material he produces and provides for other work subject to any restrictions in the Scope (Clause 70.3).

Intellectual Property Rights

The question of Intellectual Property Rights (IPR), which includes copyright, patents and trademarks, needs to be considered carefully where the Consultant is carrying out a design service.

The design and construction of a building or other structure is unique, and unless the contract stated otherwise, even though the owner pays for the design work, he does not automatically own the copyright to the design, the design documents and other materials produced by the designer.

The designer, whether it be a Consultant or a Contractor or a Supplier, retains ownership and control over the design and the various documents, based on copyright law and the terms of agreement between the parties.

This copyright issue can be very complex as it only applies to that which was originally conceived by the designer, and therefore much of what is included in the design may not be subject to copyright, for example, the layout of kitchen units or sanitary appliances within a bathroom.

Copyright also extends to copying the design, photocopying and distributing drawings and other documents, etc.

In the event that copyright is breached, the owner of the copyright may obtain an injunction preventing further breaches and, if necessary, may seek impounding of materials, and also damages, loss of profits and legal fees. If a building design copyright has been breached, the Client may be prevented from commencing or continuing with the project.

8 Liabilities and insurance

Introduction

Indemnity, insurance and liability are covered within the NEC4 Professional Service Contract by Clauses 80.1–87.1.

Insurance is, in simple terms, the principle of many paying in for the few to be compensated. It is basically a contract between the "insured" who offers to pay an agreed sum called a premium to the "insurer" who warrants to pay out an agreed sum should a particular and specified event occur, e.g. death, damage to property, etc. In effect, it is possible to insure against anything which holds an element of risk.

Types of insurance

The required insurances generally required under a construction contract can normally be grouped under the following headings.

Loss or damage to the contract works

This covers loss or damage to the construction work itself and any unfixed Plant or Materials on Site yet to be incorporated into the Works.

Public liability

This is to cover the insured against any death, injury or damage claims from members of the public other than their own Employees.

Employer's liability

In many countries, Employers of staff have a statutory obligation to provide cover for their Employees in the event of death, injury or damage caused during the course of their employment.

Professional indemnity (PI) insurance (if required)

This covers parties' liabilities particularly in respect of design and the giving of advice.

Clause 80.1 lists the Client's liabilities; however, these are the Client's liabilities only in terms of loss, wear or damage and are not an exhaustive list of all the liabilities the Clients bears under the contract.

There are many other risks such as actions or inactions of the Client or Service Manager, changes, breaches and other issues, which are covered elsewhere as compensation events.

The Parties' liabilities under the NEC4 Professional Service Contract

The Client's liabilities

The Client's liabilities are:

- claims and proceedings from Others and compensation and costs payable to Others which are due to:
  - the unavoidable result of the service or
  - negligence, breach of statutory duty or interference with any legal right by the Client or by any person employed by or contracted to it except the Consultant e.g. the Service Manager or other consultants;
- a fault of the Client or any person employed by or contracted to it, except the Consultant
- additional Client's liabilities stated in the Contract Data.

The Consultant's liabilities

The Consultant's liabilities are:

- claims and proceedings from Others and compensation and costs payable to Others which arise from or in connection with the Consultant Providing the Service,
- costs incurred by the Client which arise from a failure by the Consultant to use reasonable skill and care normally used by professional providing services similar to the service
- death or bodily injury to the employees of the Consultant.

Recovery of costs

- Any cost which the Client pays as a result of an event for which the Consultant is liable is paid by the Contractor.
- Any cost which the Consultant pays as a result of an event for which the Client is liable is paid by the Client.

Note that the right to recovery the above costs is reduced proportionally if the recovering Party is partly liable for the costs.

The Insurance Table

The Insurance Table in Clause 83.3 of the contract itemises the insurances that the Consultant has to effect, together with the *minimum amount* of cover (at least that amount). These insurances must be maintained from the starting date until the end of the periods stated in the Contract Data.

The default is that the Consultant provides, maintains and pays for the insurances, unless the Client stated otherwise in Contract Data Part 1.

The three insurances listed within the Insurance Table are as follows:

i Liability of the Consultant for claims made against it arising out of the Consultant's failure to use the skill and care normally used by professionals providing services similar to the service.

This is PI insurance (see Professional indemnity (PI) Insurance section below) and provides indemnity for any sum for which the Consultant may become liable to pay arising out of neglect, error or omission.

The amount of cover is stated in the Contract Data as being for "each claim, without limit to the number of claims"; however, most insurers set a maximum limit for any single claim and an aggregate limit to the total value of all claims per annum or for the life of the policy.

It is important to note that whilst most insurance policies deal with issues arising from action or inaction that occurred during the policy period, PI insurance policies are based on a "claims made" basis, meaning that the policy only covers claims made during the policy period when the policy is "live", so policies will usually have to be in place, and proven to be in place, long after Completion of the services to cover the Consultant's future liability.

ii Liability for damage to property and liability for bodily injury to or death of a person (not an employee of the Consultant) arising from or in connection with the Consultant Providing the Service.

This is Public Liability insurance and provides indemnity for any sum for which the Consultant may become liable to pay arising out of neglect, error or omission.

This requires the Consultant to indemnify the Client against any loss, expense, claim, etc. in respect of any personal injury or death caused by the carrying out of the work other than to their own employees. This includes the liability towards members of the public who may be affected by the Consultant's services, although they play no part in it.

In the case where a party makes a claim directly against the Client due to a death or injury, the Consultant should either take on that claim, or alternatively, the Client can sue the Consultant to recover any monies.

The Insurance Table states the minimum amount of cover or minimum limit of indemnity. The Consultant could be liable for whatever the amount of any Claim; therefore, he must consider the minimum value in the Insurance Table purely as a guide.

Note that there is a requirement for each claim, without limit to the number of claims, with cross liability so that the insurance applies to the parties separately.

The Consultant must be able to prove, if requested by the Client, that he has the required cover and his premiums are up to date, as if he does not, the Client may take out insurances himself and deduct the premiums from the Consultant.

The Consultant may also have to prove that his Subcontractors have the same cover; however, in many cases, the Consultants could either ensure that the Subcontractors have the cover or can include it under their own cover on behalf of their Subcontractors.

iii Liability for death of or bodily injury to employees of the Consultant arising out of and in the course of their employment in connection with this contract.

This is Employer's Liability insurance and provides indemnity for any sum for which the Consultant may become liable as an employer of people. This insurance provides cover against injury or death caused to people whilst carrying out their work, the requirement for such insurance being a statutory obligation in most countries (in the UK, the minimum cover has been set at £2 million).

Note that the Insurance Table refers to the greater of the amount required by the applicable law and the amount stated in the Contract Data for any one event.

The Consultant must be able to prove, if requested by the Client, that he has the required cover, and his premiums are up to date, as if he does not, the Client may take out insurances himself and deduct the premiums from the Consultant.

Again, the Insurance Table states the minimum amount of cover, the Consultant having to insure for whatever the amount of any claim.

iv Other insurances

It is possible that the Client may require the Consultant to provide other insurances e.g. for the possible replacement value of property or equipment provided by the Client for the Consultant's use. There is provision in the Contract Data for such additional insurances and also for changes to the insurances listed in Clause 81.1, for example, where the Client may himself wish to insure, for example, because he owns a building or structure within which the Consultant is to work.

There is also provision within the Contract Data for the Client to provide any of the insurances that are listed by the Consultant within the Insurance Table and also for either Party to effect additional insurances.

Insurance certificates

On request, either party must provide certificates from its insurer or broker proving that the insurances required are in force and for the required period.

Limitation of liability

If the Consultant causes, for example, any loss or damage, possibly to the Client's property, he would normally be liable for the full cost of remedial work; however, in certain situations, it may be appropriate for the Client to limit the Consultant's total liability, since unlimited liability could result in insurance premiums which would not be commercially viable, or even Consultants not wishing to tender for work without some form of limitation of liability.

Clause 87.1 provides for liability to be limited to amounts stated in the Contract Data for all matters arising under or in connection with the contract, *except* the following excluded matters for which the Consultant will retain full liability:

- Delay damages (if Option X7 applies)
- Consultant's share (if Option C applies) (this in respect of the Prices for Service Provided to Date at Completion being more than the Total of the Prices and the Consultant having to pay his share of the excess)
- An infringement by the Consultant of the rights of Others
- Loss of or damage to third-party property and death or bodily injury to a person other than an employee of the Consultant.

The Consultant is not liable for any matter unless it has been notified to the Consultant before the end of liability date which is stated in the Contract Data in terms of years after the Completion of the whole of the services.

In the UK, this may be six or twelve years depending on the type of contract, and other legislations set this at ten years.

Professional indemnity (PI) Insurance

If a Consultant providing a service to a Client makes a mistake, is found to be negligent or gives inaccurate advice, then he will be liable to the Client in the event that the Client incurs a loss as a result. This loss can be very significant where the design has to be corrected, parts of the structure have to be taken down and reinstated, a facility has to be closed down whilst the remedial measures take place, and there are also legal costs.

In such cases, it would be advisable for the Client to include a requirement for PI Insurance. PI Insurance insures the Consultant against claims for professional negligence, for instance, design defects or giving negligent advice. It is often required by professional bodies as a condition of membership.

In that event, although the Client claims against the Consultant rather than from the insurers, PI insurance protects the Consultant against claims for loss or damage made by a Client or third party.

It is important to note that PI insurance policies are based on a "claims made" basis, meaning that the policy only covers claims made during the policy period when the policy is "live", so claims which may relate to events occurring before the coverage was active may not be covered. However, these policies may have a retroactive date which can operate to provide cover for claims made during the policy period but which relates to an incident after the retroactive date. The Client should ensure that the Consultant has taken out and maintained the required PI insurances for the *full period* of his liability, which could extend to twelve years after Completion of the service.

Claims which may relate to incidents occurring before the policy was active may not be covered, although a policy may sometimes have a retroactive date, earlier than when the policy was created.

Parties taking out insurance or requiring consultants to take out insurance should pay particular amount regard to:

- the level of cover required and the excess (often referred to as a "deductible");
- whether cover relates to an (annual) aggregate amount or to each individual claim;
- whether a particular or "reputable" insurer is required. Appointments usually contain obligations relating to approved insurers of good standing and often based in the country where the work is carried out;
- exclusions from the policy;
- an obligation to maintain insurance to cover the period when claims relating to the appointment can be made; this is generally tied in to six or twelve years after completion, depending on the contract and the relevant legislation and
- the consequences of failure to maintain PI insurance: the Client might be able to put insurance in place itself and reclaim the cost from the Consultant.

PI Insurance can be applicable to costs, including defence costs arising from:

- Negligence – failing in duty of care to a Client, this can include negligent design, misstatement or misrepresentation.
- Intellectual Property – unintentional infringement of intellectual property rights, including copyright or trademark rights, including passing off others products as one's own.

- Loss of professional documents or information – this can include damage or destruction of data while in a Consultant's possession.
- Dishonesty – a dishonesty of the Consultant's partners, directors, employees or self-employed individuals directly contracted to the Consultant and under his supervision.
- Defamation – using or producing any material about other companies that may cause offence beyond the intended audience.
- Breach of confidence – inadvertent breach of confidence or misuse of any information, which is either confidential or subject to statutory restrictions.
- Any other civil liability unless specifically excluded from a policy.

It can also include liabilities arising from an Adjudicator's decision.

NB: Claims for bodily injury, property damage or personal injury will normally be excluded from a PI insurance.

Insurance of neighbouring property

Some contracts have provision for insuring neighbouring property which may suffer from collapse, subsidence, vibration, etc. due to damage by, for example, piling, deep excavations or demolition works, where not caused by the negligence of the Contractor or a Consultant.

The Professional Service Contract does not provide for this, but it can be added as a Z clause, if applicable.

The Client should consider his liabilities in respect of neighbouring owners who may be affected by the Works and any services being provided, for example, by Consultants, and, if necessary, either instruct the Contractor or Consultant to provide such insurance or take it out himself.

Such a policy would normally exclude:

- damage caused by the negligence of the Contractor, his Subcontractors or a Consultant;
- damage caused by errors or omissions in Client's design;
- damage which can be reasonably foreseen, e.g. cracking may have already occurred to a neighbouring building prior to the Works commencing – in this respect, it is important that a condition survey with photographic evidence be carried out before construction commences and
- where the damage is caused by Excepted Risks – these risks will normally be held by the Client.

9 Termination

Introduction

Termination, including reasons, termination procedures and payment are covered within the NEC4 Professional Service Contract by Clauses 90.1–93.2 with a further reference within Main Option C in respect of the Consultant's share when termination occurs.

Note also Secondary X11 Termination by the Client (see Secondary Option X11: Termination by the Client below), for a reason not included within the Termination Table.

What is termination?

Termination is a word most commonly used in the context of construction contracts to refer to the ending of the Consultant's employment. The parties have a common law right to bring the contract to an end in certain circumstances, but most standard forms give the parties additional and express rights to terminate upon the happening of specified events.

Some contracts refer to termination of the contract, whilst others refer to the termination of the Consultant's employment under the contract. In practice, it makes little difference, and most contracts make express provision for what is to happen after termination.

It should be noted that most contracts require a notice to be given by one party to the other; unless the breach is due to insolvency, the offending party then has a period of time in which to remedy the breach, failing which termination can take place.

It is important to note that neither party should attempt to terminate employment under the contract unless they are sure that the provision is available within the contract. If termination is held to be wrongful, it is usually a repudiation of the contract.

Also, termination is an entitlement not an obligation, and it should only be considered where there are no other alternatives.

Termination within the Professional Service Contract

Within the Professional Service Contract, there are twenty-two reasons (R1–R22) listed for either Party to terminate the Consultant's obligation to Provide the Service.

- Clause 91.1 (R1–R10) provides for either Party to terminate due to the insolvency of the other. This includes bankruptcy, appointment of receivers, arrangements

with creditors, winding-up orders and administration orders, which may be dependent on whether the party is an individual, a company or a partnership.
- Clause 91.2 (R11–R13) provides for the Client to terminate if the Service Manager has notified the Consultant that it has defaulted, and the Consultant has not remedied the default within four weeks of the notification of default in that the Consultant "substantially" failed to comply with its obligations R11), not provided a bond or guarantee where required (R12) or appointed a Subcontractor for substantial work before the Service Manager accepted the Subcontractor (R13).

It is quite subjective as to what the words "substantially failed to comply" and "substantial work" means within these provisions. One would assume that it is not intended to relate to minor breaches, but the Consultant could "substantially fail to comply" with a minor obligation.

- Clause 91.3 (R14–R15) provides for the Client to terminate if the Service Manager has notified the Consultant that it has defaulted and the Consultant has not remedied the default within four weeks of the notification of default in respect of the Consultant "substantially" hindering the Client or Others (R14) or "substantially" breaking a health and safety regulation (R15).

Again, it is quite subjective as to what the words "substantially hindered the Client or Others" and "substantially broken a health and safety regulation" means within these provisions. In the case of the latter, it is suggested that if the Consultant breaks a major health and safety regulation, which would probably lead to termination, but it is unclear how you can "substantially break something".

- Clause 91.4 (R16) provides for the Consultant to terminate due to non-payment by the Client within thirteen weeks of a notice from the Consultant stating that the amount is overdue. Note that the notice is a pre-condition to termination under this clause, and as with all communications, the notice being in a form which can be read, copied and recorded.
- Clause 91.5 (R17) provides for either Party to terminate if the Parties have been released under the law from further performance of the *whole* of the contract.
- Clause 91.6 (R18–R20) states that if the Service Manager has instructed the Consultant to stop or not to start any work, and the instruction to re-start has not been given within thirteen weeks.
  - the Client may terminate due to a default by the Consultant (R18).
  - the Consultant may terminate due to a default by the Client (R19).
  - either Party may terminate for any other reason (R20).
  - Clause 91.7 provides for the Client to terminate if an event occurs which stops the Consultant completing the whole of the service, or stops the Consultant completing the whole of the service by the date shown on the Accepted Programme and is forecast to delay Completion by more than thirteen weeks
  - and which neither Party could prevent, and an experienced Consultant would have judged at the Contract Date to have such a small chance of occurring that it would have been unreasonable for him to have allowed for it (R21).

    NB This clause is linked to Clause 18.1.

Procedures on termination

Under any of the twenty-two reasons, whoever terminates, the Client may complete the works, either itself or using another Consultant, and may use any Plant and Materials to which it has title (Procedure P1).

If the Client terminates for reasons R1–R15, R17, R18, R20 or R22, the Client may instruct the Consultant to assign the benefit of any subcontract to the Client (Procedure P2).

Payment on termination

Under any of the twenty-two reasons, whoever terminates the amount due is

- an amount assessed as for normal payments;
- Defined Cost incurred in reasonable expectation of completing the whole of the service;
- any amounts retained by the Client (Amount A1).

If the Client terminates for reasons R1–R15, R18–R22

The amount due on termination includes a deduction of the forecast of the additional cost to the Client of completing the whole of the service (Amount 2).

If the Consultant terminates for reasons R1–R10, R16 or R19

The amount due on termination includes the fee percentage applied to

- for Options A and C – any excess of the total of the Prices at the Contract Date and the Price for Service Provided to Date.
- for Option E – any excess of the first forecast of the Defined Cost over the Price for Work Done to Date less the Fee.

Termination and Subcontractors

The parties should also consider how termination of their contract affects Subcontractors. If the Consultant is the defaulter, termination occurs, and if the Client wishes to proceed with the works, any Subcontractors would probably be assigned to the Client.

However, if the Consultant is the terminating party, he should ensure that provisions are in place regarding termination of Subcontractors, and termination provisions in the Professional Services Contract should be mirrored in the subcontracts.

Secondary Option X11: Termination by the Client

Option X11 provides an additional reason for termination in that the Client could terminate for a reason not stated in the contract, for example, the Client has no further need for the services the Consultant is providing.

He does this by notifying the Consultant, but as the Consultant is not in breach of contract, under this option, the Client pays the Consultant 5% of the difference between the forecast final total of the Prices in the absence of termination and the total of the other amounts and costs included in the amount due on termination. This effectively gives a "loss of earnings" provision.

10 Resolving and avoiding disputes

Introduction

Resolving and avoiding disputes is covered within the NEC4 Professional Service Contract by Options W1 and W2:

- *Option W1* is selected when adjudication is the method of dispute resolution, and the United Kingdom Housing Grants, Construction and Regeneration Act 1996 *does not apply*
- *Option W2* is selected when adjudication is the method of dispute resolution, and the United Kingdom Housing Grants, Construction and Regeneration Act 1996 *applies*

Options W1 and W2 cover the adjudication process, whether the UK-based legislation, the Housing Grants, Construction and Regeneration Act 1996 is applicable (Option W2) or not applicable (Option W1).

The Act applies in England, Wales and Scotland. Similar legislation of a different title applies in Northern Ireland.

Note that the Housing Grants, Construction and Regeneration Act 1996 is supplemented by the Local Democracy, Economic Development and Construction Act 2009.

If the Professional Service Contract is used within the UK, the Client must by reference to the Act determine whether the contract is a "construction contract" and select the appropriate dispute resolution Option. If the Professional Service Contract is used outside the UK, Option W1 will always apply.

Clearly, there is nothing to prevent the parties jointly attempting to resolve a dispute as they see fit by a consensual and non-binding route such as negotiation, mediation, conciliation or mini trial, prior to referring it to adjudication.

Note that under the NEC4 Engineering and Construction Contract (but not the NEC4 Professional Service Contract), there is also an *Option W3*, which is used when a Dispute Avoidance Board is the method of dispute resolution, and the United Kingdom Housing Grants, Construction and Regeneration Act 1996 does not apply.

In recent years, contracting parties and, in turn, contract drafters have looked towards dispute avoidance in addition to dispute resolution provisions to try to prevent issues escalating into disputes in the first place. This can be in the form of more balanced contract conditions, partnering agreements, target contracts, but also by including specific dispute avoidance provisions.

Option W3 is such a provision and is a new addition to the NEC4 contracts, which is only included in, for example, the Engineering and Construction Contract.

Note that W3 is an Option for *avoiding* a dispute, *not resolving* it.

A potential dispute under the contract may be referred to the Dispute Avoidance Board, two to four weeks after notification to the other Party and to the Project Manager.

The Parties are required to make all relevant information available to the Dispute Avoidance Board, who visits the Site (where applicable) and helps the Parties to settle the potential dispute *without* formal referral as a dispute.

What is adjudication?

Adjudication has been defined as "a summary non-judicial dispute resolution procedure that leads to a decision by an independent person that is temporarily binding but which may subsequently be reviewed by means of arbitration, litigation or by agreement".

It is therefore an intermediate dispute resolution process which is binding upon the parties unless and until the matter is referred to a tribunal, normally arbitration or litigation.

In the UK, adjudication is a statutory right, but it has always been included as the primary means to resolve a dispute within the NEC contracts, failing which the parties may refer a dispute to a tribunal. In fact, the majority of adjudication decisions is accepted by parties as the final result and is never taken to the tribunal.

There have been debates over the years regarding whether adjudication is appropriate for major, complex and/or high value disputes or for disputes which are not related to payments, for example, negligent design.

The author has some sympathy with that view; however, under the NEC contracts, a Party cannot refer a dispute to the tribunal (normally arbitration or legal proceedings) *unless* the dispute has first been referred to the Adjudicator so this could be a burden where adjudication is felt by *both* Parties not to be the best way to resolve the dispute.

Statutory right to adjudication (UK-based contracts)

When the Housing Grants Construction and Regeneration Act 1996 was introduced in the UK, it applied to construction contracts which came into existence after 1 May 1998, and apart from a few notable exceptions, it provided a statutory right to refer a dispute under a construction contract to adjudication. This Act has now been amended by the Local Democracy, Economic Development and Construction Act 2009.

The Act applies to all what it terms as "construction contracts". That term captures many activities that feature in service contracts even though construction in its normal definition might not be involved, including:

- all normal building and civil engineering work, including construction, alteration, repair, maintenance, extension and demolition or dismantling of structures forming part of the land and works forming part of the land, whether they are permanent or not;
- the installation of mechanical, electrical and heating works and maintenance of such works;

- cleaning carried out in the course of construction, alteration, repair, extension, painting and decorating and preparatory works;
- agreements with consultants such as architects, other designers, engineers and surveyors;
- elements such as scaffolding, site clearance, painting and decorating;
- labour-only contracts and
- contracts of any value.

It excludes:

- work on process plant and on its supporting or access steelwork on sites where the primary activity is nuclear processing, power generation, water or effluent treatment, handling of chemicals, pharmaceuticals, oil, gas, steel or food and drink;
- supply-only contracts, that is contracts for the manufacture or delivery to site of goods where the contract does not provide for their installation;
- extracting natural gas, oil and minerals (and the workings for them);
- purely artistic work;
- off-site manufacture;
- contracts with residential occupiers. This means a contract where work is carried out on a dwelling which one of the parties occupies or intends to occupy as his residence rather than simply a housing project;
- PFI contracts. This only includes the head agreements in PFI projects, not a contract for the construction of the works and
- finance agreements. This includes loan agreements.

The Adjudicator is appointed jointly by both parties who are jointly and severally liable for those fees.

Under the Act, the contract must include an adjudication clause which complies with the requirements of the Act. If the contractual clause does not comply, then it is voidable, and the parties can exercise their right to adjudication under the 'Scheme for Construction Contracts' issued by the Secretary of State.

The provisions in the contract must:

- enable the parties to give notice at any time;
- provide a timetable with the object of securing the appointment of the adjudicate and referral of the dispute to him within seven working days;
- require the Adjudicator to reach a decision within twenty-eight days of referral or such longer period as is agreed by the parties after the dispute has been referred;
- allow the Adjudicator to extend the period of twenty-eight days by up to fourteen days with the consent of the party by whom the dispute was referred;
- impose a duty on the Adjudicator to be impartial;
- enable the Adjudicator to take the initiative in ascertaining the facts and the law;
- provide that the decision of the Adjudicator is binding until determined by litigation, arbitration or by agreement and
- provide that the Adjudicator is not liable for anything done or omitted in the discharge or purported discharge of his functions as Adjudicator unless the act or omission is in bad faith, and that any employee or agent of the Adjudicator is similarly protected from liability.

Who can be an adjudicator?

Whilst many professional bodies hold registers of Adjudicators, there is no formal qualification to become an Adjudicator, though there are commonly acceptable requirements which include the following:

- A recognised professional qualification in a relevant discipline. The relevant discipline then relates to the dispute in question, for example, if the dispute is related to services installations, a Mechanical or Electrical Engineer would probably be a relevant discipline, though it must be remembered that the Adjudicator's role is to enforce the contract, so in that case, he does not necessarily have to be a Mechanical or Electrical Engineer just because the dispute is related to services. Clearly, as disputes are always of a legal nature, lawyers tend to make up a good proportion of the adjudicating community.
- Substantial number of years' experience within the industry, normally ten to fifteen years post qualification in a relevant discipline.
- A qualification, or at least a thorough understanding of contracts and legal principles, though in many cases, the Adjudicator is not a lawyer, nor does he have to be. Many say that it is better if he is not a lawyer, but that is probably a view left to the cynics.
- Recognised training in the law and practice of adjudication.
- Membership of a professional institution.
- Some also say that the Adjudicator should have Professional Indemnity (PI) insurance, though it must be emphasised that as the Adjudicator is not liable to the Parties for any action or inaction on his part other than in bad faith, it could be questioned as to whether such insurance should be a prerequisite to appointment. Most adjudicator nominating bodies require their panel members to carry appropriate insurance.

As the use of adjudication in the resolution of contractual disputes has grown, particularly in the UK where parties have a statutory right to refer a dispute to adjudication, many education establishments and professional bodies now provide diploma and certificate courses in adjudication.

Who pays for the adjudication?

The cost of the Adjudicator's fee is allocated by him/her between the parties, the Adjudicator deciding what the proportion is to be paid by each party, with in some cases, one party paying the majority or even all of the Adjudicator's fee. However, both parties are jointly and severally liable for payment of the Adjudicator's fee; therefore, if one party will not or cannot pay his share, possibly due to insolvency, the other party must pay it and recover the amount directly from the defaulting party.

Each party bears his own costs in professional fees, collating, copying, postage, etc. which is different to, for example, litigation where both parties' costs are borne by the losing party.

What will the adjudicator charge?

There are no set hourly rates for Adjudicators, each Adjudicator will advise the party of his hourly rate and whether there are other costs payable, for example, for travel, accommodation or printing. The Adjudicator is required to properly record the time spent carrying out his duties.

What if a party does not comply with the adjudicator's decision?

If a party does not comply with the Adjudicator's decision, the other party can enforce that decision by referring to court.

Selecting the adjudicator

Within Contract Data Part 1, there are three options for selecting the Adjudicator:

i The name and address of the Adjudicator may be stated by the Client in Contract Data Part 1, and he is then appointed before the starting date.

This has the advantage that tendering Consultants are aware at the time of tender who will be the Adjudicator in the event that a dispute arises and, if necessary, can address the issue and object to them within their tender submissions. Also, the Adjudicator is already in place should a dispute arise. The named Adjudicator may require some form of retainer fee for being named in the contract and being available in the event that a dispute arises.

ii The parties can mutually agree to the name of the Adjudicator in the event that a dispute arises. This has the advantage that there is no "Adjudicator in waiting" and therefore no fee payable. Many Professional Service Contract practitioners say that pre-appointing the Adjudicator as in option (i) is resigning oneself to the fact that there will at some point be a dispute. However, once parties are in dispute, they then have to agree who is to be the Adjudicator and appoint him, though, at this point, they may not wish to agree with each other about anything.

iii The name of the Adjudicator nominating body may be stated by the Client in Contract Data Part 1. This is probably the favoured option as an independent name can be put forward by the nominating body, who is normally well prepared to offer the name of an Adjudicator from its registers and have him appointed within the time scales set by the contract and if necessary the appropriate legislation.

In all cases, the Adjudicator must be impartial i.e. the Adjudicator should not show any bias towards either party. In addition, all correspondence from the Adjudicator must be circulated to both parties.

Any request for an Adjudicator must be accompanied by a copy of the Notice of Adjudication, and the appointment of the Adjudicator should take place within seven days of the submission of the Notice of Adjudication to the other party.

Adjudicator nominating bodies

Adjudicator Nominating Bodies, as the name suggests, are organisations that fulfil the role of nominating Adjudicators. These bodies keep registers of Adjudicators with varying expertise and based at various geographical locations who can act for parties should they be nominated.

Clearly, in naming the Adjudicator Nominating Body in the contract, it is advisable to name an organisation with expertise in the project to be carried out.

These bodies are very knowledgeable about appointment of Adjudicators, and the relevant time scales and for a modest fee can nominate a suitably qualified Adjudicator to suit the parties' requirements.

Preparing for adjudication

Once the adjudication process starts by the issuing of a Notice of Adjudication by the referring party, it is a very quick procedure; therefore, it is critical that the referring party and, in turn, the responding party prepare well in the short time they have available. Time scales are stated within the contract, though these may be extended by agreement between the parties.

As the process does not start until the referring party initiates the process by issuing the Notice of Adjudication to the other party, the referrer does have the advantage of time, and in many cases surprise, as he can prepare his case before issuing the notification.

Before considering initiating the adjudication process, parties should consider the following:

i There must be a dispute – whilst this may seem fairly obvious, it is essential that a dispute actually exists between the parties in that one party has made a claim, which the other party denies (probably repeatedly) and that the parties have exhausted the means to resolve the dispute themselves, for example, by the use of negotiation, mediation, conciliation, etc., though they do not have to prove that they have tried other methods.

ii The dispute arises under or in connection with the contract – the Adjudicator is appointed to decide a dispute under the contract; therefore, it is essential that the dispute arises on a matter covered by the contract. If the dispute is whether a contract exists in the first place, then an Adjudicator cannot decide that.

iii The role of the Adjudicator is to enforce the contract not to decide who is the most deserving party. It is therefore essential that, before commencing adjudication, that the referring party ensures that he has properly complied with the contract; otherwise, he may find himself in a weak position.

There have been adjudications in the past where a Consultant has acted in good faith but not strictly in accordance with the contract, accepting and willingly acting on verbal instructions in order not to damage what appeared to be a good relationship with the Client, then when a dispute arises, the Adjudicator states that as the contract did not provide for verbal instructions, the Consultant was incorrect to act on them; and therefore, the Consultant loses the adjudication.

iv Whilst many people say that, as adjudication is a fairly simple process, it should not include lawyers as the issues are industry-related rather than legal issues, it is important that the parties carry out their obligations within the short time scale they have, so if a party is uncertain as to what they have to do, and when and how they have to do it, they obtain professional advice.

v It is essential that a clear and concise notice of adjudication is prepared by the referring party as it defines what is in dispute and the matters that are being referred to the Adjudicator. The Adjudicator cannot consider issues which are not set out within the Notice of Adjudication.

The notice should include:

- Brief details of the service.

Whilst the Adjudicator may have been named in the contract, he may have little knowledge about the actual project.

- The names and addresses of the parties to the contract and the parties in dispute.
- The nature and a brief description of the dispute.
- The nature of the redress which is sought from the Adjudicator.

It is critical that the responding party, and also the Adjudicator, clearly understands what redress the referring party is seeking by referring the matter to adjudication.

i Details included within the referral should be clearly and concisely presented so that the responding party knows what it needs to respond to, and also the Adjudicator clearly understands what is required of him. Any supporting information should be clearly referenced and relative to the matter in dispute.

  Neither party should not include or rely on any information or evidence which the other party has not seen. If they do, the other party may raise an objection, which could bring the adjudication process to a premature end without a resolution.

ii The contract provides for time scales to be extended by agreement between the parties, so if more time is needed in order to prepare and submit documentation, the other party should be consulted.

Does a "dispute" exist?

Only a dispute between the contracting parties can be referred to adjudication. If there is no dispute, then an Adjudicator has no jurisdiction to consider the matter.

Whilst this seems obvious statement, it is a critical question to ask, "has the matter crystallised into a dispute"?

As an example, if the Consultant notifies a compensation event and the Service Manager does not agree that it is a compensation event, then the simple fact that he does not agree does not mean that there is a dispute as the parties are still free to meet and, if necessary, to negotiate an outcome. It is only when the parties have exhausted *all* the opportunities to meet and reach a compromise, but still have a difference that a dispute exists, the matter can be referred to the Adjudicator.

It must also be noted that an Adjudicator only has jurisdiction to consider one dispute at a time under the same contract unless there is an agreement between the disputing parties. Therefore, if there are number of issues in dispute, unless they are inextricably joined, then there is likely to be several adjudications.

Disputes under or in connection with the contract may be referred to the Adjudicator.

Note that a dispute under the contract cannot be referred to the tribunal (arbitration or litigation) unless it has first been decided by the Adjudicator.

Adjudication is a quick, convenient and relatively inexpensive way of resolving a dispute, whereby an impartial third-party Adjudicator either named in the contract, selected by agreement between the parties or nominated by a third party, decides the issues in dispute between the parties. It is quicker and less expensive than arbitration or litigation, though the parties may refer to one of those methods following adjudication.

It must be recognised that the role of the Adjudicator is to enforce the contract, not to decide what, in his opinion, would be the fairest outcome. For example, if the parties have a dispute over the quality of the replacement of suspended ceiling tiles carried out, it is the Adjudicator's task to ascertain whether the tiles were supplied and installed in accordance with the contract, i.e. specifications, etc. and the applicable law, not whether it was a good or bad job. In addition, the question of whether a contract exists cannot be decided by the Adjudicator but more likely through the courts.

Dispute resolution provisions within the Professional Service Contract

As stated above, the Professional Service Contract provides two Options for referral to the Adjudicator.

Option W1

Option W1 is used when the UK Housing Grants, Construction and Regeneration Act 1996 does not apply.

Any dispute arising under the contract is first referred by a Party issuing a notice to the Senior Representatives (who are named in the Contract Data Parts 1 and 2 on behalf of the Client and the Consultant, respectively) and also copied to the other Party and the Service Manager, stating the nature of the dispute it wishes them to resolve.

Note that a Party may replace a Senior Representative after notifying the other Party of the name of the replacement.

Each Party then submits their statement of case (no more than ten sides of A4 paper) with supporting evidence.

The Senior Representatives then *try* to resolve the dispute within three weeks, using any procedure they wish to (they have to *try*, but they are *not* obliged to resolve the Dispute). At the end of the three weeks, a list is issued showing the issues which may have been agreed, and the issues which may not have been agreed. The Service Manager and the Consultant put the agreed issues into effect.

The Dispute Resolution Table includes the following:

If the Dispute is about:

- an action or inaction of the Service Manager, *either Party* may refer it to the Senior Representatives, not more than four weeks after the Party became aware of the action or inaction.
- a programme, compensation event or quotation for a compensation event which is treated as having been accepted, *the Client* may refer it to the Senior Representatives, not more than four weeks after it was treated as accepted.
- an assessment of Defined Cost which is treated as correct, *either Party* may refer it to the Senior Representatives, not more than four weeks after the assessment was treated as correct.
- any other matter, *either Party* may refer it to the Senior Representatives, when the dispute arises.

Note that if the referring Party is the Client and the dispute is regarding a compensation event which has been treated as having been accepted (Clause 62.6), then the Client must notify the dispute to the Consultant and then refer it to the Adjudicator not more than four weeks after it was treated as accepted.

This seems an odd reason for referring a dispute to the Adjudicator, as Clause 62.6 states that if the Service Manager does not reply to the Consultant's notification, it is treated as acceptance of the quotation, how could an Adjudicator whose role is to enforce the contract find other than in favour of the Consultant? Surely, the Client's dissatisfaction, and the dispute itself is, with his Service Manager, not with the Consultant?

Whilst these time scales are fixed within the Adjudication Table, they can be extended by the Service Manager if the Consultant and the Service Manager agree before either the notice or the referral is due. Note that if the matter in dispute is not notified and referred within these time scales, neither Party can refer the dispute to the Adjudicator or the tribunal.

The Adjudicator is appointed under the NEC4 Dispute Resolution Service Contract. The Adjudicator acts impartially, and if he resigns or is unable to act, the Parties jointly appoint a new Adjudicator. The referring Party obtains a copy and completes the Dispute Resolution Service Contract. If the Parties have not appointed an Adjudicator, either party may ask the nominating body to choose an Adjudicator.

No procedures have been specified for appointing a suitable person, and in practice, a number of different methods have been used. Whatever method is used, it is important that both Parties have full confidence in his impartiality, and for that reason, it is preferable that a joint appointment is made.

The Adjudicator should be a person with experience in the type of work included in the contract between the Parties and who occupies or has occupied a senior position dealing with disputes. He should be able to listen and to understand the viewpoint of both Parties.

Often, the Parties delay selecting an Adjudicator until a dispute has arisen, although this frequently results in a disagreement over who should be the Adjudicator.

As noted, the selection of the Adjudicator is important, and it should be recognised that a failure to agree an Adjudicator means that a third party will make the selection without necessarily consulting the Parties.

The referring Party must include within his referral information which he wishes to be considered by the Adjudicator, any more information from either Party to be provided within four weeks of the referral.

A dispute under a subcontract, which is also a dispute under the Professional Service Contract, may be referred to the Adjudicator at the same time, and the Adjudicator can decide the two disputes together.

The Adjudicator may review and revise any action or inaction of the Service Manager, take the initiative in ascertaining the facts, and the law relating to the dispute, instruct a party to provide further information and instruct a party to take any other action which he considers necessary to reach his decision.

All communications between a Party and the Adjudicator must be communicated to the other Party at the same time.

The Adjudicator decides the dispute and notifies the Parties and the Service Manager within four weeks of the end of the period for receiving information.

This four-week period may be extended by joint agreement between the Parties and the Adjudicator.

Until this decision has been communicated, the Parties proceed as if the matter in dispute was not disputed.

The Adjudicator's decision is temporarily binding on the parties unless and until revised by a tribunal and is enforceable as a contractual obligation on the parties.

The Adjudicator's decision is final and binding if neither Party has notified the Adjudicator that they are dissatisfied with an Adjudicator's decision within the time stated in the contract and intends to refer the matter to the tribunal.

The Adjudicator may, within two weeks of giving his decision to the Parties, correct a clerical mistake or ambiguity.

Review by the tribunal

A dispute *cannot* be referred to the tribunal unless it has first been referred to the Adjudicator.

The tribunal may be named by the Client within Contract Data Part 1, whilst no alternatives are stated, it will normally be litigation or arbitration.

If arbitration is chosen, the Client must also state in Contract Data Part 1, the procedure, the place where the arbitration is to be held, and the person or organisation who will choose an Arbitrator if the Parties cannot agree a choice, or if the named procedure does not state who selects the Arbitrator.

A Party can, following the adjudication, notify the other Party within four weeks of the Adjudicator's decision that he is dissatisfied. This is a time barred right, as the dissatisfied Party cannot refer the dispute to the tribunal unless it is notified within four weeks of the Adjudicator's decision, failure to do so will make the Adjudicator's decision final and binding.

Also, if the Adjudicator has not notified his decision within the time provided by the contract, a party may within four weeks of when the Adjudicator should have given his decisions notify the other party that he intends to refer the dispute to the tribunal.

The tribunal settles the dispute and has the power to reconsider any decision of the Adjudicator and review and revise any action or inaction of the Service Manager.

It is important to note that the tribunal is not a direct appeal against the Adjudicator's decision, and the parties have the opportunity to present further information or evidence that was not originally presented to the Adjudicator, and also the Adjudicator cannot be called as a witness.

Option W2

Option W2 is used when the UK Housing Grants, Construction and Regeneration Act 1996 applies.

The inclusion of adjudication within the then New Engineering Contract, now the NEC4, pre-dates UK legislation giving parties to a contract the statutory right to refer a dispute to adjudication.

The Local Democracy, Economic Development and Construction Act 2009 which came into force in England and Wales on 1 October 2011 and in Scotland on 1 November 2011 amended the Housing Grants, Construction and Regeneration Act 1996, the new Act applying to most UK contracts after that date.

There still remains certain categories of contract to which neither Act applies.

The differences between the Housing Grants, Construction and Regeneration Act 1996 and the Local Democracy, Economic Development and Construction Act 2009 include:

- the previous Act only applied to contracts which were in writing, and the new Act also applies to oral contracts.
- terms in contracts such as 'the fees and expenses of the Adjudicator as well as the reasonable expenses of the other party shall be the responsibility of the party making the reference to the Adjudicator' will be prohibited. Much has been written about the effectiveness of such clauses, often referred to as Tolent clauses (Bridgeway Construction Ltd vs. Tolent Construction Ltd 2000) and whether they comply with the previous Act, so the new Act should provide clarity for the future.
- the Adjudicator is permitted to correct his decision so as to remove a clerical or typographical error arising by accident or omission. Previously, he could not make this correction.
- Section 108 of the Act provides parties to construction contracts with a right to refer disputes arising under the contract to adjudication. It sets out certain minimum procedural requirements which enable either party to a dispute to refer the matter to an independent party who is then required to make a decision within twenty-eight days of the matter being referred.

If a construction contract does not comply with the requirements of the Act, or if the contract does not include an adjudication procedure, a statutory default scheme, called the Scheme for Construction Contracts (referred to as the 'Scheme'), will apply.

The Act provides that a dispute can be referred to adjudication "at any time", provided the parties have a contract.

"At any time" can also refer to a dispute after the contract is completed.

What are the requirements within construction contracts?

Section 108 of the Act requires all "construction contracts", as defined by the Act, to include minimum procedural requirements which enable the parties to a contract to give notice at any time of an intention to refer a dispute to an Adjudicator.

The contract must provide a timetable so that the Adjudicator can be appointed, and the dispute referred, within seven days of the notice. The Adjudicator is required to reach a decision within twenty-eight days of the referral, plus any agreed extension, and must act impartially. In reaching a decision, an Adjudicator has wide powers to take the initiative to ascertain the facts and law related to the dispute.

The Housing Grants, Construction and Regeneration Act 1996 defines adjudication as:

> a summary non judicial dispute resolution procedure that leads to a decision by an independent person that is, unless otherwise agreed, binding upon the parties for the duration of the contract, but which may subsequently be reviewed by means of arbitration, litigation or by agreement.

In that sense, adjudication does not necessarily achieve final settlement of a dispute because either of the parties has the right to have the same dispute heard afresh in court or where the contract specifies arbitration.

However, experience since the Housing Grants, Construction and Regeneration Act 1996 came into force shows that the majority of adjudication decisions are accepted by the parties as the final result.

For Option W2, there is no Adjudication Table as the Parties have a right to refer a dispute to each other and to the Adjudicator at any time.

The remainder follows generally as with Option W1.

Any dispute arising under the contract is first referred by a Party issuing a notice to the Senior Representatives and copied to the other Party and the Service Manager stating the nature of the dispute it wishes them to resolve.

Note that a Party may replace a Senior Representative after notifying the other Party of the name of the replacement.

Each Party then submits their statement of case (no more than ten sides of A4 paper) with supporting evidence.

The Senior Representatives then try to resolve the dispute within three weeks, using any procedure they wish to. At the end of the three weeks, a list is issued showing the issues agreed and the issues not agreed. The Service Manager and the Consultant put the agreed issues into effect.

The requirements of the Housing Grants Construction and Regeneration Act 1996 for adjudication to be possible at any time mean that the involvement of the Senior Representatives may be bypassed by parties and the dispute referred straight to the adjudicator.

Again, as with Option W1, the Adjudicator is appointed under the NEC4 Dispute Resolution Service Contract. The Adjudicator acts impartially, and if he resigns or is unable to act, the Parties jointly appoint a new Adjudicator. The referring Party obtains a copy and completes the Dispute Resolution Service Contract. If the Parties have not appointed an Adjudicator, either party may ask the nominating body to choose an Adjudicator.

A Party may first give a notice of adjudication to the other party with a brief description of the dispute, details of where and when the dispute has arisen, and the nature of the redress which is sought.

The Adjudicator may be named in the contract in which case the party sends a copy of the notice to the Adjudicator. The Adjudicator must confirm within three days of receipt of the notice that he is able to decide the dispute, or if he is unable to decide the dispute.

Within seven days of the issue of the notice of adjudication, the party:

- refers the dispute to the Adjudicator;
- provides the Adjudicator with the information on which he relies together with supporting information and
- provides a copy of the information and supporting documents to the other party.

Again, a dispute under a subcontract, which is also a dispute under the Professional Service Contract, may be referred to the Adjudicator at the same time, and the Adjudicator can decide the two disputes together.

The Adjudicator may review and revise any action or inaction of the Service Manager, take the initiative in ascertaining the facts, and the law relating to the dispute, instruct a party to provide further information and instruct a party to take any other action which he considers necessary to reach his decision.

All communications between a Party and the Adjudicator must be communicated to the other party at the same time.

The Adjudicator decides the dispute and notifies the parties and the Service Manager within 28 days of the dispute being referred.

This twenty-eight-day period may be extended by fourteen days with the consent of the referring Party or by any other period by joint agreement between the parties.

Until this decision has been communicated, the parties proceed as if the matter in dispute was not disputed.

The Adjudicator's decision is temporarily binding on the parties unless and until revised by a tribunal and is enforceable as a contractual obligation on the parties. The Adjudicator's decision is final and binding if neither party has notified the Adjudicator that they are dissatisfied with an Adjudicator's decision within the time stated in the contract and intends to refer the matter to the tribunal.

The Adjudicator may, within five days of giving his decision to the parties, correct any clerical mistake or ambiguity.

Review by the tribunal

A dispute cannot be referred to the tribunal unless it has first been referred to the Adjudicator.

Again, the tribunal may be named by the Client within Contract Data Part 1, normally litigation or arbitration. If arbitration is chosen, the Client must also state in Contract Data Part 1, the procedure, the place where the arbitration is to be held and the person or organisation who will choose an Arbitrator if the Parties cannot agree a choice or if the named procedure does not state who selects the Arbitrator.

A party can, following the adjudication, notify the other party within four weeks of the Adjudicator's decision that he is dissatisfied. This is a time-barred right, as the dissatisfied Party cannot refer the dispute to the tribunal unless it is notified within four weeks of the Adjudicator's decision, failure to do so will make the Adjudicator's decision final and binding.

The tribunal settles the dispute and has the power to reconsider any decision of the Adjudicator and review and revise any action or inaction of the Service Manager.

It is important to note that the tribunal is not a direct appeal against the Adjudicator's decision, and the parties have the opportunity to present further information or evidence that was not originally presented to the Adjudicator, and also the Adjudicator cannot be called as a witness.

NEC4 Dispute Resolution Service Contract

Introduction

The NEC4 Dispute Resolution Service Contract replaced the NEC3 Adjudicator's Contract and was launched in June 2017.

The first edition of the NEC Adjudicator's Contract was published in 1994 and was written for the appointment of an Adjudicator for any contract under the NEC family of standard contracts. The second edition, published in 1998, contained some changes including the need to harmonise with the NEC standard contracts and further editions which had been issued since 1994. The third edition harmonised the contract with the NEC3 family of contracts using either Option W1 or W2.

The Dispute Resolution Service Contract is made up of five parts:

1 General
2 Adjudication
3 Dispute Avoidance Board
4 Payment
5 Termination

The Dispute Resolution Service Contract can also be used with contracts other than NEC4, though dependent on the contract and the applicable law, some amendment may be necessary.

In the UK, the Housing Grants, Construction and Regeneration Act 1996 has made adjudication mandatory as a means of resolving disputes in construction contracts which fall under the Act. Parties to a contract which does not provide for adjudication as required by the Act have a right to adjudication under the "Scheme for Construction Contracts". Schemes which are substantially similar have been published for England and Wales, Scotland and Northern Ireland.

The agreement is between the two disputing parties and the Dispute Resolver, who may also be acting as the Adjudicator.

The contract contains five sections:

1 General

   This covers the obligation upon the Dispute Resolver to act impartially and to notify the parties as soon as he becomes aware of any matter that could present a conflict of interest.

   There is a definition of expenses, which includes printing costs, postage, travel, accommodation and the cost of any assistance with the adjudication.

   All communications must be in a form that can be read, copied and recorded. This is a requirement of all NEC contracts, but also of the adjudication process itself, with both Parties and the Dispute Resolver having to be copied into any communication between the parties.

2 Adjudication

   This Clause only applies if the Dispute Resolver is acting as an Adjudicator.

   The Dispute Resolver cannot decide any dispute that is the same or substantially the same as his predecessor decided. He must make his decision and notify the Parties in accordance with the contract, and in reaching his decisions, he can obtain assistance from others, but must advise the Parties before doing so and also provide the Parties with a copy of anything produced by the assisting party, so that the Parties can be invited to comment.

   The Dispute Resolver's decision is to remain confidential between the parties, and following his decision, the Dispute Resolver retains documents provided to him for the period of retention, which is stated in the Contract Data.

   An invoice is issued by the Dispute Resolver:

   - each time a dispute is referred if an advance payment is stated in the Contract Data;
   - after a decision had been notified to the Parties and
   - after a termination.

3 Dispute Avoidance Board

This Clause only applies if the Dispute Resolver is acting as a Dispute Avoidance Board member.

The Dispute Resolver carried out the duties of a Dispute Avoidance Board member in accordance with the contract between the Parties and collaborates with other members of the Dispute Avoidance Board.

4 Payment

The Dispute Resolver's hourly fee is stated in the Contract Data. There is provision for an advance payment to be made by the referring party to the Dispute Resolver if stated in the Contract Data.

Following his decision, the Dispute Resolver submits an invoice to each Party for their share of the amount due, including expenses. The Parties pay the amount due within three weeks of receiving the Dispute Resolver's invoice. Interest is paid at the rate stated in the Contract Data on any late payment. The Parties are jointly and severally liable to pay the amount due to the Dispute Resolver; therefore, if one party fails to pay, the other must pay including any interest on the overdue amount and recover the amount from the other Party.

5 Termination

The Parties may, by agreement, terminate the Dispute Resolver's agreement. Also, the Dispute Resolver may terminate the agreement if there is a conflict of interest, he is unable to decide a dispute, an advance payment has not been made or he has not been paid an amount due within five weeks of the date by which the payment should have been made. The Dispute Resolver's appointment terminates on the date stated in the Contract Data.

The tribunal

Whilst Contract Data Part 2 requires the Client to enter the form of the tribunal, it does not say what choices are available. Other contracts refer disputes to expert determination or Dispute Review Boards, but under the Professional Service Contract, the Client normally selects arbitration or litigation.

Within the Contract Data, the Client is required to state the arbitration procedure, the place where the arbitration is to be held, and who will choose an Arbitrator.

There are a number of differences between the two but essentially, arbitration is hearing the dispute in private, and litigation is hearing the dispute in a public court.

It must be stressed that the role and authority of the tribunal is not to review and appeal against the previous decision of an Adjudicator, but to resolve the dispute independent of the adjudication; therefore, it is quite in order for parties to introduce new evidence or materials to the tribunal.

Arbitration

Any dispute between two or more parties can be resolved through the courts, a process known as litigation.

However, litigation has many disadvantages, not least the cost of a full court hearing, but, in some cases, a lengthy waiting time before the matter actually reaches the courts, so arbitration has become increasingly used as a simpler, more convenient method of dispute resolution.

Arbitration is the hearing of a dispute by a third party, who is often not a lawyer, though that term has a very wide meaning, but is an expert in the field in which the dispute is based, and can give a decision based on opinion, which is then legally binding on both parties.

Arbitration is not a new concept; in fact, it has been in existence for almost as long as the law itself, the first official recognition in the UK being the Arbitration Act 1697, which largely governed disputes about the sale of livestock in cattle markets. With the continuing growth of freight transport by ship, and the advent of railways and other forms of transport, a large number of arbitration cases ensued, and as a result, Parliament passed the Arbitration Act 1889.

Various amendments have been introduced over the following years with the current legislation being the Arbitration Act 1950, as amended by the Arbitration Acts of 1975 and 1979. The Arbitration Act 1996 is the current legislation related to arbitration in England and Wales.

One does not issue writs in arbitration as one would with litigation, *both* parties agree to enter into arbitration as a pre-agreed provision within a contract and thus be bound by the decision of the Arbitrator.

Arbitrators are appointed by one of THREE methods:

i As a result of a pre-agreed decision between the parties, normally though a provision within a contract to go to arbitration in the event of a dispute.

   In this case, the Arbitrator himself may be named or subject to nomination by a professional body such as the Chartered Institute of Arbitrators.

ii By Statute, legislation in many countries includes the provision for disputes to be settled by arbitration.

iii By Order of a Court.

The Arbitration Act is fairly brief and sets out procedures in the absence of any agreement to the contrary between the parties.

As arbitration is a more flexible arrangement than litigation, whatever procedure the Arbitrator and both parties agree is sufficient in an arbitration case.

Whilst most forms of contract provide for resolution of disputes by arbitration, it must be stressed that it should only be seen as a last resort when all attempts at negotiation and other resolution methods have failed.

It may be surprising to hear that arbitration is common as a settlement procedure for disputes under trade agreements, maritime and insurance, consumer matters such as package holidays, construction industry disputes and property valuations.

The contracting party that initiated that the dispute be referred to arbitration is always referred to as the 'Claimant', whilst the other party is referred to as the 'Respondent'. Should there be a Joint Agreement to go to arbitration, the Arbitrator will decide who is the Claimant and who is the Respondent.

It is important that these titles are clarified as the Arbitration Rules repeatedly refer to them.

There are several *advantages* in using arbitration rather than litigation:

i Privacy

   This is a distinct advantage, as in litigation, private and often personal matters are debated in full view of the public in open court and often published in the press and technical magazines, whereas arbitration is carried out in private sessions, usually at the Arbitrator's office or at a pre-arranged venue.

ii Convenience

Arbitration is carried out at a time and place to suit the parties, for example, evenings and weekends, not when the court is available for session.

iii Speed

The dispute is settled efficiently without the normal delays and procedural matters involved in often fairly lengthy and formal litigation.

iv Simplicity

There are few technical procedures as in the courts, and the parties set their own procedures within a set of ground rules.

v Expertness

The Arbitrator will be an expert on the matter in dispute, for example, a structural engineer where the dispute is about structural design issues, whereas judges and other legal professional are experts only on principles of law. The dispute will often not hinge on a legal problem, for example, in matters of design who better than a designer to resolve it?

vi Thoroughness

Because the Arbitrator is a skilled technical person, the dispute can often be dealt with more thoroughly and matters more rigorously aired than in a confined and formal court atmosphere.

vii Expense

It can be readily seen that arbitration is generally a less expensive option than litigation, though it must not be seen as a cheap way to resolve a dispute. In some complex cases, arbitration can be even more expensive than litigation.

It is worth mentioning at this point that, in many countries, legal aid, which can be available for litigation, is not available for arbitration.

The main *disadvantages* of using arbitration rather than litigation are:

i Lack of legal knowledge

A dispute may hinge on a difficult point of law, which the Arbitrator has insufficient legal knowledge to resolve, as he is an expert in the field relating to the dispute, not in the law itself. He is, however, entitled to seek legal advice on a point of law, but not on the dispute itself, and also the disputing parties can refer a question of law to the High Court.

NB The arbitration award itself cannot be referred to the Courts unless both parties consent or the Courts are referred to on a point of law, the resolution of which could substantially overturn an Arbitrator's decision.

Further, Arbitrators cannot be sued for negligence as they do not hold themselves up as experts in the sense of being advisory bodies and therefore liable for giving wrongful advice.

ii Precedents

Arbitration is not subject to rulings in previous cases as with the courts, each case being judged on its own merits on the day, and also to some extent subject to the personality of the Arbitrator; therefore, there is no guide as to how successful an action could be.

It is prudent to always seek professional advice before resorting to arbitration as whilst it is simpler than Court action, the Arbitrator's decision is final and binding; therefore, a party must be confident that they have a strong case.

iii Decision-making

It has been said that as the Arbitrator, being a technical person and possibly having been involved in a similar dispute themselves, tends to be fairer than judges in litigation, they also tend to sympathise with both parties and can split the difference between the parties when making their awards as they have possibly been in a similar position themselves, but judges are inclined to take colder decisions and award wholly in favour of one of the parties. It has to be said that this is not a commonly held view.

Reference to arbitration

Although arbitration is a different process to adjudication, the same initial principles apply before arbitration can commence in that there must be a dispute, there must be an agreement to arbitrate, which will normally be provided for within the contract, and third, a party must submit a notice to initiate the process.

Proceedings begin by either party sending a written notice to the other stating that they wish to refer the dispute to arbitration, the form and process may be dictated by the arbitration procedure stated by the Client within Contract Data Part 1.

The parties may agree on a person to act as Arbitrator, or a professional body will select and appoint the Arbitrator. In some cases, where the parties cannot agree a choice, and there is no provision for a third party to do it, a court may select an Arbitrator. There is normally only one Arbitrator, but in some cases, three, consisting of two Arbitrators and a third who acts as an umpire. Any correspondence flowing between the two parties as from the date of appointment must then also be copied to the Arbitrator(s).

The next step is that the Arbitrator writes to the parties accepting the appointment, a Preliminary Meeting is held between the parties at a time and venue to be directed by the Arbitrator to decide the form of the hearing which may be on a documents only basis or an oral hearing. Documents only tend to be cleaner and quicker as there is less opportunity for the parties to debate the issue.

The role of the arbitrator

The role of the Arbitrator is only to examine the evidence before him enabling him to resolve the dispute, and nothing more, he cannot look into other matters not in dispute. He can conduct the arbitration as he wishes, though he must gear the procedures in a way that is not detrimental to either party.

He must also have a regard for the rules of evidence.

For example:

1 Hearsay evidence is not permissible.
2 He is not permitted to use evidence from someone not qualified to give such evidence.
3 The Arbitrator is not allowed to take into account any evidence he has discovered himself, which was not provided by either party.
4 He is not allowed to use any privileged evidence, for example, that which is marked 'Without Prejudice'.
5 In addition, the rules of discovery apply where either party is entitled to have access to the file or documents of the other provided that it relates to matters at issue.

Inspection by the arbitrator

The Arbitrator is entitled to inspect any property, work or material at any premises and may invite the Claimant and/or the Respondent to accompany him purely for the purpose of identifying the work or materials.

No-one else is allowed to accompany the Arbitrator unless he specifically invites him.

The award

The award is binding upon both parties and must be served in the appropriate manner with full headings, summary of issues in dispute, outline of events leading to the dispute, his decision based on the hearing and finally the monetary award.

There is very limited scope for appeal against an Arbitrator's award. Usually, appeals can only be based on a claim that the Arbitrator was wrong on a point of law or there was serious and proven irregularity.

Arbitration awards can also be enforced in court if necessary.

Arbitrator's fees and expenses

Both parties will normally incur costs, though it is the Arbitrator's decision as to how the costs are apportioned. Both parties are initially liable to the Arbitrator for payment of his fees and expenses, but the Arbitrator will direct the parties as to the proportion they are liable for once the decision is reached.

Should the parties agree on a settlement of the dispute at any time during the arbitration, they will jointly be liable for the Arbitrator's costs.

Litigation

It is not proposed to go into any detail about the litigation process from initiation to judgement by a court, first, as the contract does not give any details of the process, but also the proceedings themselves will depend on the law of the contract, which, in turn, is dependent on the location of the project.

However, the process will normally involve the Claimant's representative, usually, a Solicitor issuing a claim form or writ to the appropriate court, papers are then served on the Defendant. The Defendant then replies to the service by either admitting or denying the claim against him; if he denies it, he replies by sending his defence which may also be accompanied by any counter-claim.

The case is then allocated to a court. There then follows a period of accumulation, collation and presentation of evidence, and hearings, again the process, the submission and the form of the hearing being dependent upon the law of the contract, until the court passes judgement.

Cases in most UK courts will normally see barristers acting as advocates on behalf of the parties alongside the solicitors who have advised the parties.

11 Tenders

Deciding the procurement strategy

Deciding the procurement strategy is the foundation stone to a successful outcome to a contract for providing professional services, but this process is very often understated or almost overlooked as Clients and their advisers, in their urgency to get projects out to tender, and their Consultants appointed tend to select the procurement strategy and standard form of contract most familiar to them rather than to spend time properly considering the project in terms of the Client's resources and his objectives in terms of time, price and quality, with the additional factor of risk having to also be considered and evaluated.

It is rarely achievable for all three to be fully satisfied, and it is usually a balance (see Figure 11.1).

The initial briefing with the Client is the fundamental first step in establishing what is required, when it is required and what the Client is proposing and/or can afford to pay. Through a process of interviewing key members of the Client's team and reviewing his business plans, specific objectives, options and constraints, a procurement strategy is developed which forms the basis for future assessments and recommendations.

Whilst it may sound obvious, it is vital that the Client understands what he wants and can also communicate it clearly. The Client may not know what he wants, but also he can often be swayed by Consultants or other advisers or third parties into something he does not want.

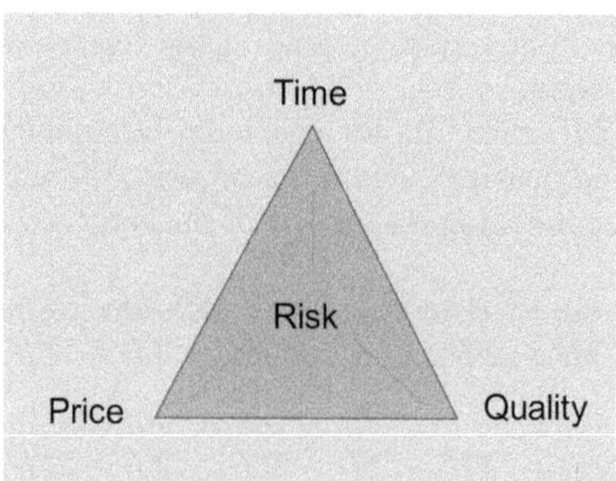

Figure 11.1 Time – Price – Quality – Risk triangle.

In broad terms, the procurement strategy should consider time, price and quality and the balance between the three as follows:

i Time

Most Clients want professional services to be completed as quickly as possible.

As an example, if the professional service is design, the Client wants to get the Contractor on board as soon as possible to build what the Consultant has designed.

This is particularly relevant in sectors such as retail, where the Client needs to open a shop as early as possible in order to sell goods. Also, a Client may need to tie in completion of a project with transfer of staff, students or patients from older buildings. Also, some projects have absolute deadlines e.g. international sporting events.

If the Client sees time as being the most important factor, he may have to pay more money, and quality may have to be lowered in order to achieve that. The procurement method will have to consider timing through programme requirements, but also through provisions such as sectional completion and delay damages.

ii Price

All Clients will have some form of budget, so this will always be a major consideration. The budget may be fixed, representing all that the Client has available through his own funds or borrowings in order to fund the project or may have some degree of flexibility. The form of procurement, ranging from lump sum through to time based, will need to address the budget and any required flexibility.

iii Quality

Finally, quality will always be a major factor, though if the above two are more important, many Clients will expect to lower standards either by not having what they want or by value engineering, which considers similar but more economical solutions.

It is important that the designer lowers the standards by agreement, not that a "cheap" Consultant is employed who will unilaterally lower the standards himself. It is a lot more difficult to detect if a Consultant has lowered standards than if a Contractor has done so.

More detailed questions can then be asked and answered to finally complete the procurement strategy, which include the following:

- How important to the Client is certainty of price, particularly if changes are expected? Lump sum options will give greater certainty of price, and time-based options will give less certainty of price, but greater flexibility.
- What resources and expertise does the Client have? Time-based and target contracts require resources to check invoices, time sheets, etc. as opposed to lump sum contracts. How important to the Client is an early start and/or early completion? Early commencement and early completion may lead the Client towards a time-based solution as lump sum and re-measurement procurement options require a considerable time to prepare the tender documentation for pricing.
- How clearly definable is the Scope, and what form does it take, for example, is it in the form of detailed requirements or performance-based criteria?
- What is the likelihood of change to those defined requirements? Whichever option is selected, the NEC4 Professional Service Contract has a dynamic change management system (compensation events) which allows changes to be priced speedily.

- What is the Client's attitude towards risk? Who is best placed to manage the inherent risks within the project? Lump sum contracts assign greatest risk to the Consultant, and time-based contracts assign greater risk to the Client.

It is a poorly acknowledged fact that the Client always pays for risk, regardless of whether it is carried within the contract by the Client or Consultant. If it is the Consultant's risk, he is deemed to have priced and programmed for it, and if it is a Client's risk, he will pay for it if it occurs.

The procurement strategy then leads to a contract strategy which considers what is available in terms of contract to enable the Client to achieve the objectives drawn from the procurement strategy. Within the Professional Service Contract, appropriate decisions can then be made through the main and secondary options.

Partnering Arrangements

The past fifteen years, particularly, since the publication of the Latham and Egan reports, have seen the growth of partnering and framework agreements.

The US Construction Industry Institute has defined partnering as:

> A long-term commitment between two or more organisations for the purpose of achieving specific business objectives by maximising the effectiveness of each participant's resources... the relationship is based upon trust, dedication to common goals and an understanding of each other's individual expectations and values.

Partnering is a medium to long-term relationship between contracting parties, whereby Consultants and various other parties are not required to tender competitively on price for each project, but are awarded the work on the basis of their ability to deliver and normally by price negotiation.

The construction industry is known to be a high-risk business, and many projects can suffer unexpected cost, and time overruns frequently resulting in disputes between the parties. The risks within a project are initially owned by the Client, who may choose to adopt a "risk transfer" approach where the risks are assigned through the contract to the Consultants who have the opportunity to price and programme for them, or a "risk embrace" approach where the Client retains the risks. In reality, most contracts are a combination of the two. The traditional approach to risk management is that of risk transfer, which is fine if the scope of work is clear and well defined; however, in recent years, Clients have become more aware that they can achieve their objectives better by adopting a more "old fashioned" risk embrace culture.

Advantages of partnering agreements are as follows:

- The procurement process from conception to commencement on site, and subsequent completion can be significantly reduced in terms of time and resources.
- No re-tendering time and cost for future projects.
- Relationships can be developed based on trust.
- Consultants are appointed earlier and can contribute to the design and procurement process.
- There tends to be greater cost certainty.

- Continuous improvement can be achieved by transferring learning from one project to the next.
- Better working relationships can be developed as the parties know each other.
- Prices can be more competitive and resources used more efficiently by continuous flows of work.

Disadvantages of partnering agreements are:

- Obligations and liabilities can become less clear in time as the parties can lapse into informal working patterns.
- Complacency can set in after some time as the Consultants have an assured flow of work.
- Clients are often unsatisfied that they are getting value for money in their projects.
- Continuing to award the work to a small number, or even one Consultant, prevents other equally as good or better providers having the opportunity to carry out work.

All of these disadvantages can be overcome by maintaining a disciplined approach to communications between the parties and their rights and obligations and also by continuous measurement of performance and deliverables.

Invitation to Tender

As soon as it has been decided to select a Consultant by competitive tender rather than by negotiation, a shortlist of those considered suitable to be invited to tender should be compiled. Sometimes, expressions of interest, interviews and presentations can be considered prior to drawing up the list.

It is critical that all the Consultants on the shortlist have the proven competence, resources, health and safety record, and financial stability to carry out the works, so that when tenders are received and assessed, no doubt is in the Client's mind as to whether a preferred Consultant has the ability and resources to carry out the project. Clients should also review their shortlists periodically so as to exclude firms whose performance has been proven to be unsatisfactory and also to allow the introduction of new Consultants.

The cost of preparing tenders is a significant element of the overheads of the Client, the Consultants and, in turn, their Subcontractors and Suppliers, so lists of tenderers should be kept as short as practicable. Depending on the size and type of project, it is recommended that the shortlist should be from four to six names.

EU Procurement Directives

Whilst not wishing this book to be geography-specific, it is worth some consideration of the detailed European Union (EU) procurement directives that apply to all procurement within the public sector, above minimum monetary thresholds which are reviewed on a regular basis, and subject to EU wide principles of non-discrimination, equal treatment and transparency.

The purpose of the procurement rules is to open up the public procurement market and to ensure the free movement of supplies, services and works within the EU.

The rules are enforced by the Members States' Courts and the European Court of Justice and require that all public procurement must be based on value for money, defined as "the optimum combination of whole-life cost and quality to meet the user's requirement", which should be achieved through competition, unless there are compelling reasons to the contrary.

In addition to the EU Member States, the benefits of the EU public procurement rules also apply to a number of other countries outside Europe because of an international agreement negotiated by the World Trade Organization (WTO) titled the "Government Procurement Agreement" (GPA).

The EU Procurement Directives set out the legal framework for public procurement when public authorities and utilities seek to acquire supplies, services or works (e.g. civil engineering or building), procedures which must be followed before awarding a contract when its value exceeds set thresholds which are reviewed on a regular basis, unless the contract qualifies for a specific exclusion, for example, on grounds of national security.

The Directives require contracting authorities to provide details of their proposed procurement in a prescribed format, which are then published in the *Official Journal of the European Union* (*OJEU*). All companies replying to an *OJEU* advertisement have an equal opportunity to express interest in being considered for tendering. As in all tendering exercises, Clients must ensure that those companies selected to tender receive exactly the same information on which to make their bid.

The notice may be issued electronically, and this service may be provided through an intermediary organisation.

Generally, contracts covered by the Regulations must be the subject of a call for competition by publishing a Contract Notice in the *OJEU*.

Choice of procurement procedure

Four-award procedures are provided in the Regulations:

- the *open procedure*, where all those interested may respond to the advertisement in the *OJEU* by tendering for the contract;
- the *restricted procedure*, under which a selection is made of those who respond to the advertisement, and only they are invited to submit a tender for the contract. This allows purchasers to avoid having to deal with an overwhelmingly large number of tenders;
- the *competitive dialogue procedure*, following an *OJEU* Contract Notice and a selection process, the authority then enters into dialogue with potential bidders, to develop one or more suitable solutions for its requirements and on which chosen bidders will be invited to tender;
- the *negotiated procedure*, under which a purchaser may select one or more potential bidders with whom to negotiate the terms of the contract. An advertisement in the *OJEU* is usually required but, in certain circumstances, described in the Regulations, the contract does not have.

The *OJEU* Notice process has associated minimum timeframes for receipt of tenders or requests to participate, depending on procurement procedure, from the date the contract notice is sent.

Stages in the procurement process

The Regulations set out criteria designed to ensure all suppliers or contractors established in countries covered by the rules are treated on equal terms to avoid discrimination on the grounds of origin in a particular Member State.

The criteria cover:

- Specification stage – how requirements must be specified.
- Selection stage – the rejection or selection of candidates.
- Award stage – the award of contract, either on the basis of "lowest price" or preferably the "most economically advantageous tender" to the purchaser.

Preparing tender documents

The tender documents for a Professional Service Contract will normally consist of:

i Invitation to Tender (including instructions to tenderers on time and place of tender submission).

There is no "pro forma" version of "Instructions to Tenderers" within the NEC contracts, but any invitation to tender will give a brief description of the work, who it is for and how tenders may be submitted.

Other inclusions will cover such matters as:

- time, date and place for the delivery of tenders;
- what documents must be included with the tender, including a programme;
- the policy regarding on alternative and/or non-compliant bids;
- arrangements for visiting the site including contact details;
- rules on non-compliant bids and
- anti-collusion certificate.

ii Form of Tender

There is no "pro forma" Form of Tender within the NEC contracts, though the Professional Service Contract Guidance Notes include a sample form. This is the tenderer's written offer to execute the work in accordance with the contract documents.

The Form of Tender is normally in the form of a letter with blank spaces for tenderers to insert their name and other particulars, total tender price and other particulars of their offer. It is essential to have a standard Form of Tender and that all tenderers consistently use the same form to make comparison of tenders easier.

iii The Pricing Document, possibly a pro forma activity schedule for an Option A or C contract.

iv The Scope

The Client provides the Scope and refers to in Contract Data Part 1.

> The Scope is defined under Clause 11.2(13) as:
> Information which either

- specifies and describes the service or
- states any constraints as to how the Consultant Provides the Service

and is either

- in the documents which the Contract Data states it is in or
- in an instruction given in accordance with the contract

The Scope is the equivalent of "Works Information" and to a certain degree "Site Information" in an NEC3 Engineering and Construction Contract and will include a detailed description of the service to be provided, but also any constraints on how the service is provided e.g. working hours, access to places, etc. It should also include how success or failure in Providing the Service is to be measured.

The main documents within the Scope could include the following:

Description of the service

- What is the Consultant being appointed to do? What are the deliverables?
- What is required to be submitted, when and in what format?

Information

- What information is being provided to the Consultant? If work is required on existing land, or within an existing building or structure, any available existing schematic layouts, plans, elevations and section drawings, etc.
- Depending on what the Consultant is appointed to do, there could be other information about existing land, buildings or structures including:
  - ground investigations, borehole and trial pit records and test results;
  - details of any previously demolished structures and the likelihood of any residual surface and subsurface materials;
  - reports obtained by the Client concerning the physical conditions within the Site or its surroundings; this may include mapping, hydrographical and hydrological information;
  - information on the topography of the site;
  - environmental issues, for example, nesting birds and protected species;
  - references to publicly available information and
  - information from utilities companies and historic records regarding plant, pipes, cables and other services below the surface of the site.

Constraints

- A statement of any specific constraints on how the Consultant Provides the Service e.g. working hours, restrictions on access, sequencing or phasing of works and security issues.
- Specific health and safety requirements which the Consultant is required to comply with, again, particularly if he is required to work within existing premises, including house and local safety rules, evacuation procedures, etc.

Facilities and services provided by the Client or Others

- Anything that will be provided by the Client e.g. manpower, technical support and advice, temporary accommodation, use of IT.

  Also whether a charge will be applied for providing the facilities or services and how the charge will be levied. The Consultant has the right to use anything provided by the Client, but only to Provide the Service.
- Any pre construction information and health and safety plans for the project.

Materials provided by the Consultant

- The Client has the right to use material provided by the Consultant for the purpose stated in the Scope. The Consultant is also required to obtain from any Subconsultant equivalent rights for the Client and for himself (see also Option X9).

Payment

- Details required to be shown in invoices submitted by the Consultant.
- Details of the accounts and records to be kept by the Consultant.

Acceptance Procedures

- Any acceptance procedures should be included including time scales for submission and acceptance.

Completion

- Completion is defined under Clause 11.2(2), but the Scope should also define any specific requirements which are required in order for completion to have taken place, for example, and requirement for:
  - "as built" drawings,
  - maintenance manuals,
  - training documentation,
  - test certificates and
  - statutory requirements and/or certification.

Documents to be retained by the Consultant after Completion

- The Scope must identify any documents to be retained by the Consultants after Completion, including the form in which they should be kept, e.g. hard and/or electronic copies and the pace of storage. This can include copies of drawings, specifications, correspondence, photographs, reports, etc. which record the services.

  These must be retained by the Consultant for the period of retention stated within Contract Data Part 1, normally six or twelve years (Clause 13.9).

Other parties

- Details of other Consultants, Contractors and Others who will be working in the same location during the contract period and any sharing requirements.

Subcontracting

- Lists of acceptable Subcontractors for specific tasks
- Statement of any work which should not be subcontracted
- Statement of any work which is required to be subcontracted

Quality Management System

- Details of the quality management system, quality policy statement and quality plan to be provided by the Consultant (Clauses 40.1 and 40.2).

Programme

- Any information which the Consultant is required to include in the programme in addition to the information shown in Clause 31.2. Also, if the Client requires the Consultant to produce a certain type of programme or to submit it using a certain brand of software, then this should be clearly detailed within the Scope.

Others

- There are also certain specific requirements for statements to be made in the Scope from certain main and secondary options in the conditions of contract.
  - X4.1 the form of any ultimate holding company guarantee.
  - X13.1 the form of any performance bond.

The Scope must be carefully drafted in order to define clearly what is expected of the Consultant in the performance of the contract and therefore included in the quoted tender amount and programme. If the Scope does not cover all aspects of the work, either specifically or by implication, that aspect may be deemed to be excluded from the contract.

A comprehensive, all embracing description should therefore be considered for the Scope, which should be supplemented by specific detailed requirements. If reliance is placed solely on a very detailed Scope description, an item may be missed from this detailed description and be the subject of later contention.

The Scope describes clear boundaries for the service to be provided by the Consultant. It may also outline the Client's objectives and explain why the work is being undertaken and how it is intended to be used. It says what is to be done (and maybe what is not included) in general terms, but not how to do it or the standards to be achieved. It explains the limits, where the work is to interface with other existing or proposed facilities. It may draw attention to any work or materials to be provided by the Client or others. It should also emphasise any unusual features of the services, which tenderers might otherwise overlook.

This is the document that a tendering Consultant can look to, to gain a broad understanding of the scale and complexity of the job and be able to judge its capacity to undertake it. It is written specifically for each contract. In some respect, it is analogous to a shopping list. It should be comprehensive, but it should be made clear that it is not intended to include all the detail, which is contained in the drawings, specifications and schedules.

Again, it may lay down the order in which various portions of the services are to be executed. As far as possible, it should describe the outcomes required rather than how to achieve them.

The Scope describes what the Consultant has to do and, in some cases, must not do or include in order to Provide the Services. It may also include the order or sequence in which the works are to be carried out. It also details where the work is to interface with other existing or proposed Consultants or Contractors. It will also include any work or facilities or materials to be provided by the Client or others. It may also include any unusual features of the work, which tenderers might otherwise overlook, for example, any planning constraints, etc.

This is the document that from which a tender can gain a broad understanding of the scale and complexity of the job and his capacity to undertake it.

It is written specifically for each contract.

The Scope should be:

- Clear – Unambiguous
- Concise – Not over wordy
- Complete – Have nothing missing

Example of a Scope

A Consultant is to be appointed to carry out structural tests in respect of existing railway bridge structures.

The Scope for this testing must fully define the full extent and timing of tests and inspections required, where the tests are to be carried out, what format should the results take and what must be done in the event that an irregularity or problem is identified.

The compiler of the Scope should consider the following questions:

a What is to be tested?

What elements or components are required to be tested? Is it to be tested in an assembled or unassembled state?

b What is the purpose of the tests?

What is it intended to prove or disprove? Is a particular test required to be carried out before something else can happen? Is payment to the Consultant dependent on successful Completion of the test?

c How are the tests to be carried out and who is to provide the Materials, Facilities and Equipment to do the tests?

Whether the tests are a well-known "industry standard" test or not, the exact nature of the test must be clearly defined. Who supplies the Equipment, who pays for the power or fuel for the Equipment, etc.? Is a component to be tested to destruction? It is also important to detail who is to provide the Materials and Facilities to do the test and who meets the costs.

d Who should do the tests? Who should be present and who should be invited should they wish to attend when the tests are taking place?

There may be a requirement for an independent party to carry out part of the test or to be present to witness the test and to provide results.

e When and where will the tests take place?

Clearly state when the tests must take place. Sometimes, tests have to be carried out on a periodic basis.

f What is the expected result or outcome from the tests?

What are we expecting the tests to prove? What results do we need to gather once the tests have been carried out?

g Who should be advised of the outcome of the tests?

The Client will normally be advised of the outcome of a test, but sometimes, a third party outside the contract may need to be informed.

h What should be done in the event of a successful test e.g. certification?

Does a certificate have to be issued following successful Completion of a test?

Does that successful outcome mean that the Consultant can proceed to the next stage of the work?

i What should be done in the event of an unsuccessful test, e.g. rectification or replacement, and further testing?

Does the Consultant have to repeat the test? Is the retest done immediately after the first test, or should there be a period before the next test? Is it possible that even though something has failed a test, it may still be considered acceptable?

j What about additional testing, i.e. to confirm a suspected Defect or the extent of a suspected Defect?

v Contract Data Part 1 (completed by the Client; see below "Contract Data Part 1");

vi Contract Data Part 2 (blank pro forma to be completed by tendering Consultants; see below "Contract Data Part 2").

In addition, there may be other information, dependent on the applicable legislation, for example, in the UK, where the Construction (Design & Management) Regulations 2007 apply, Pre Construction Information would be included within the tender documents.

The Contract Data

The Contract Data provides the information required by the conditions of contract specific to a particular contract. Other contracts call this the "appendices" or "contract particulars".

Contract Data part 1

Contract Data Part 1 consists of data provided by the Client, the sections of the Contract Data aligning with the sections of the core clauses.

Section 1 – General

This section requires the Client or the party representing the Client to identify the selected main and secondary options, the names of the Client, the Service Manager and the Adjudicator.

The "service" and the "Scope" are also identified within this section, though these are normally incorporated by reference to separate documents rather than listing drawing numbers and specification references. If this is the case, the separate documents must be clearly defined.

The service can be simply a couple of words of description such as "Quantity Surveying Services" or "Carrying out Traffic Survey", the detail of which is shown in the Scope, which specifies and describes the services and any constraint on how the Consultant Provides the Services.

As the NEC contracts are intended for use worldwide, the language and the law of the contract are also entered.

The period for reply is the period that the parties have to reply to submissions, proposals, notifications, etc. where there is no period *specifically* stated elsewhere within the contract, for example, the period within which a party should reply to an early warning notice, or the Client should reply to the Consultant submitting an early warning notice, or the name of a proposed Subcontractor. The period of reply would clearly not apply to the Consultant's submission of a programme for acceptance or the submission of a quotation for a compensation event.

The period for retention of documents is then stated as a number of years following Completion or termination, whichever is earlier. Dependent on the law of the contract, this could be a period of, for example, six or twelve years.

With regard to dispute resolution, the Adjudicator nominating body is named within this section, normally, a professional institution, and then the tribunal is named, which is normally either arbitration or legal proceedings.

Finally, within this section, any matters to be included in the Risk Register will be identified.

Section 2 – The Consultant's main responsibilities

This section starts with identification of any Key Dates and Conditions to be met.

It also allows the Client to insert a period when, under Option A, forecasts of the total expenses, and, under Options C and E, the forecasts of the total Defined Cost plus Fee have to be prepared, usually monthly.

Section 3 – Time

The starting date is inserted within this section, also details of any access which will be provided, the requirement for revised programmes, the completion date for the while of the service, and the period after the Contract Date when the Consultant is to submit a first programme for acceptance.

Section 4 – Quality management

The Client states the timing for the submission of a quality policy statement and quality plan as a period of weeks from the Contract Date, the Contract Date being the date the contract comes into existence.

The defects date is identified as a number of weeks after Completion of the whole of the services. This identifies the period the Consultant is initially liable for correcting Defects, though, again dependent on the law of the contract, he may be liable for correcting Defects for a longer period after.

Section 5 – Payment

The assessment interval is entered, which is normally identified as "one calendar month" rather than in "weeks".

Again, as the NEC contracts are intended for use worldwide, the currency of the contract is entered.

The Client states the interest rate per annum above a designated rate of a designated bank. The interest rate is applicable to late payment and is assessed from the date when the payment should have been made, until the date when it is made. This interest is calculated on a daily basis and is compounded annually.

Section 6 – Compensation events

This section is only completed if there are additional compensation events.

Note that under the Engineering and Construction Contract, the Client is required to insert details of weather information and the measurement of weather conditions, in respect of a potential weather-related compensation event; however, there is no weather-related compensation event under the Professional Service Contract, but if the contract was to be amended to provide for weather events, then weather information will be required within this section.

Section 7 – Title

This section is not included within the Contract Data as no entries are required.

Section 8 – Liabilities and insurance

Any additional Client's liabilities are listed here, over and above those listed within Clause 80.1.

The minimum limit of indemnity for Professional Indemnity (PI) Insurance, Public Liability and the Consultant's liability for his own employees (Employer's Liability Insurance) is stated within this section.

Also, any additional insurances, and any insurances that may be provided by the Client.

Resolving and avoiding disputes

The Client enters details of the choice of tribunal, normally arbitration or legal proceedings, also if arbitration is selected, the arbitration procedure, where the arbitration is to be held, and who will choose an Arbitrator if the Parties cannot agree a choice.

The Senior Representatives on behalf of the Client are also named and the choice of Adjudicator (if named), see also Chapter 10 for more details.

Secondary options

Depending on which Secondary Options are selected, the Client enters details for the relevant entry in the Contract Data.

Option X1: Price adjustment for inflation
Option X2: Changes in the law
Option X3: Multiple currencies
Option X5: Sectional Completion
Option X6: Bonus for early Completion
Option X7: Delay damages
Option X8: Undertakings to Others
Option X10: Information Modelling
Option X11: Termination by the Client
Option X12: Multiparty collaboration
Option X13: Performance Bond
Option X18: Limitation of Liability
Option X20: Key Performance Indicators
Option Y(UK)1: Project Bank Account
Option Y(UK)2: Housing Grants, Construction & Regeneration Act 1996
Option Y(UK)3: The Contracts (Rights of Third Parties) Act 1999
Option Z: Additional conditions of contract

Contract Data part 2

Contract Data Part 2 is completed by the Consultant as part of his tender and includes:

Section 1 – General

- The name and address of the Consultant is inserted.
- The fee percentage applicable to Defined Cost is inserted here.
- The names of key people, their name, job, responsibilities, qualifications and experience.
- Any matters to be included within the Early Warning Register (see Chapter 2) may be added by the Consultant.

Section 2 – The Consultants' main responsibilities

Any Scope provided by the Consultant is inserted here.

Section 3 – Time

- The reference details for the programme submitted with the Consultant's tender are inserted here.
- The Client may decide the Completion Date or may invite tendering Consultants to do so.

Section 5 – Payment

- Any expenses the Consultant requires are stated here.
- The reference details for the activity schedule submitted with the Consultant's tender are inserted here.

Resolving and avoiding disputes

The Senior Representatives on behalf of the Consultant are named here.

Secondary options

Option X10 (Information Modelling), the Consultant identifies the information execution plan (if required).

Submission of tenders

There is a tendency to give Consultants very short periods of time in which to submit their tenders as the Client wishes the work to be commenced as soon as possible; however, it is critical that tenderers are given sufficient time in which to properly consider what they have to price, and time and resource-related issues, and to make any enquiries and obtain quotations prior to submitting their tenders.

The time and place for the delivery of tenders needs to be clearly stated. It is also critical that all tenderers are treated on an equal basis; therefore, it should be made clear to all tenderers that late returns will be invalid and will not be considered. Late tenders should be returned to the tenderer unopened.

The tender documentation should also make clear the period over which a tender may be required to remain open for acceptance. This would normally be three months, giving the Client and his advisers time to properly assess the tenders and make the appropriate award.

Opening tenders

Again, in order to maintain equality between the tenderers, tenders should not be opened until after the date and time for submission to avoid any suspicion of collusion. No tenderer should be allowed to alter his tender after the closing time.

After opening, the tenders and supporting documents should be evaluated. Each tender is confidential, and no details of a tender should be released to any other tenderer.

Notification of tender results

It is important that all tendering Consultants be informed of the results of tendering as soon as possible after the tender submission date. This particularly applies to those whose tenders are not successful, or which are not being given further consideration, as it frees them to concentrate on opportunities with other Clients.

Generally, Clients' representatives have no authority to unilaterally accept tenders, as ultimately, the contract is between the Client and the Consultant, so it is normal to recommend a choice to the Client and allow them the final decision and inform the tenderers.

The Clients' representatives should submit a report, which lists the tenders, received, comments on them and recommends a particular one for acceptance. The scope and detail of this report will vary according to the circumstances.

The report should present a clear and logically reasoned case for acceptance of the recommended tender.

The following may be included in the report:

- A tabular statement of all the tenders received, showing the name of each tenderer, the tender sum, whether it was a valid submission and any qualifying conditions.
- Reference to any discussions that may have taken place with any tenderers regarding the tender.
- A concise summary of the findings following the examination and assessment of each tender, reasons for considering any tender invalid, any discussion of the programme and methods proposed and reasons for considering any of these to be unsatisfactory.
- Comments on any rates or prices, which appear exceptionally high or low, and forecasts of the possible effects on the contract price in cases where there are large differences.
- A comparison of the tender of the recommended successful bidder with the original cost plan or budget.
- Recommendation of the most acceptable tender.

Remember – the cheapest price does not always mean the cheapest cost!

As discussed earlier, by carrying out a detailed pre-tender analysis, when tenders are received, the evaluation criteria should, in theory, be fairly straightforward as all the tenders should be capable of being accepted, so it will normally just include any further information gained from the bid itself.

Awarding the contract

There is no Standard Form of Agreement for completion and signature within the NEC4 contracts, though within the Professional Service Contract Guidance Notes, there is a sample form. The drafters of the contract have stated that parties enter into contracts in many different ways, by executing pro forma agreements, by exchange of letters, etc. Whilst this is probably true, it would have assisted the parties if there had been a standard Form of Agreement that they could use if they so wish.

The difficulty with not having a Form of Agreement within the Contract is that the parties have to write their own, they also have to consider whether the contract is to be executed as a deed, and in that respect, they must be very cautious that all the documents that are intended to form the contract are included in the Form of Agreement.

12 The NEC4 Professional Service Short Contract

Why a short form of contract?

The majority of building and engineering contracts and for that matter consultancy contracts carried out are of relatively low risk and complexity and often but not necessarily, low value.

Hence the work involved in preparing a full NEC4 Professional Service Contract may not be justified in many cases.

Also, the detailed procedures and management systems in the Professional Service Contract may not be necessary in many contracts.

Thus, there was a definite need for a shorter contract which is simple, easy to use and more suited to the low risk and complexity, and small value type of contract.

The need for shorter contracts has been recognised by most of the contract writing bodies.

For example, JCT publishes the "Minor Works Building Contract", ACE/CECA the "ICC Infrastructure Conditions of Contract Minor Works Version" and FIDIC the "Short Form of Contract (The 'Green Book')".

The Latham Report *Constructing the Team* emphasised the need for a shorter contract, paragraph 5.20 of the report stating that "provision should be made for a simpler and shorter minor works document" and a footnote states "for example, the Landscape Institute regards the NEC as too complex even for landscape services in connection with civil engineering projects".

The first page of the NEC4 Professional Service Short Contract states:

> This contract should be used for the appointment of a supplier to provide a professional service which does not require sophisticated management techniques, comprises straightforward work and impose only low risks on both client and consultant.

Unlike other forms of contract, there is no limiting or recommended financial value for contracts under the Professional Service Short Contract, because the above criteria are not in general related to a particular sum of money. It is possible to have large value, low risk and simple work carried out under the Professional Service Short Contract. On the other hand, it may not be appropriate to carry out complex high risk work under the Professional Service Short Contract even though its monetary value is small.

By contrast, the many Minor Works contracts are normally recommended for work not exceeding £250,000 in value with contract periods not exceeding six months. In practice, however, these forms are often used for work in excess of the recommended values.

Intended users of the Professional Service Short Contract range from major organisations, carrying out a large number of straightforward, low-risk projects, to domestic householders seeking a simple, user-friendly contract to enable them to appoint a consultant for a home extension. It has to be said that the great majority of users of the contract fall into the former category.

The history of the Professional Service Short Contract is as follows:

- First edition 2013
- Second edition June 2017

Since the Professional Service Short Contract has been drafted for use on straightforward and low-risk work, many matters covered by the Professional Service Contract have been omitted.

Whilst the intention has been to cater for anything which might go wrong on the simpler project, it is always possible that something may occur which is not covered in the Professional Service Short Contract.

Such matters will then have to be dealt with by the law of the contract.

A comparison between the Professional Service Contract and the Professional Service Short Contract is given below.

Arrangement of the Professional Service Short Contract

The Professional Service Short Contract includes the following sections:

Contract Forms

- Contract Data
- The Consultant's Offer and Client's Acceptance
- Price List
- Scope

Conditions of Contract

1 General
2 The Consultant's main responsibilities
3 Time
4 Quality management
5 Payment
6 Compensation events
7 Rights to material
8 Liabilities and insurance
9 Termination and resolving disputes

The roles of the parties

There are three parties referred to in the Professional Service Short Contract:

- the Client
- the Consultant
- the Adjudicator

The Client

The Parties to the contract are the Client and the Consultant.

The Client is rarely mentioned in most construction contracts, the Project Manager (ECC), Engineer, Architect or some other party representing him, but in the Professional Service Short Contract, the Client represents himself, though he will usually delegate to someone to represent him (see Client's Representative below).

Actions that cannot be delegated by the Client are:

i Allowing access and use of the site, as the site belongs to the Client.
ii Making payment to the Consultant, as the Client is the other Party to the contract.

The Consultant

The Consultant and the Client are the Parties to the contract, the Consultant's responsibility being to Provide the Service in accordance with the contract, the Scope describing the service and any constraints on how he has to provide it.

There are also obligations in respect of complying with instructions, notifying early warnings, submitting programmes for acceptance, and notifying and submitting quotations for compensation events.

The Client's Representative

Unlike the NEC4 Professional Service Contract, there is no definite "Service Manager" role under the NEC4 Professional Service Short Contract, all actions being directly between the Client and the Consultant, though under Clause 14.5, the Client, after notifying the Consultant, may delegate any actions and may cancel any delegation.

Though the term "Client's Representative" is not stated anywhere in the contract, someone may be appointed by the Client, with the authority of the Client delegated to him to manage the contract on his behalf.

A Client's Representative may be appointed from the Client's own staff or may be an external consultant.

If a consultant, they may be appointed using the NEC4 Professional Service Short Contract though this is not mandatory. The Client's Representative is a named individual and not a company.

It is vital that the Client appoints a representative who has the necessary knowledge, skills and experience to carry out the role, which includes acceptance of programmes (where necessary), issuing and responding to early warnings, certifying payments and dealing with compensation events.

Disciplines which have carried out the role to date include Engineers from a civil engineering, structural or process background, Project Managers, Architects, Building Surveyors and Quantity Surveyors.

It is also vital that the Client gives any representative full authority to act for him, particularly when considering the time scales imposed by the contract.

If he has to seek approvals and consents from the Client, then he must do so, and the appropriate authority must be given, in compliance with the contractual time scales.

If the approval process is likely to be lengthy, it is critical that both the Client and his representative set up an accelerated process to comply with the contract or that the time scales in the contract are amended. The former is the vastly preferred method in order that NEC4 will work to its full effect.

Although the Client's Representative manages the contract at the post-contract stage, he is usually appointed pre-contract to deal with matters such as feasibility issues, advising on design, procurement, cost-planning tendering and programme matters.

Note that, unlike many other contracts, the NEC contracts do not name the Quantity Surveyor, so this would need to be considered in dealing with financial matters pre- and post-contract. Either the Client or his representative may carry out the role of the Quantity Surveyor himself or delegate to another.

As with any other notification under the contract, if the Client wishes to delegate, the notice to the Consultant must be in writing i.e. by letter or email, not by verbal communication such as a telephone call.

The delegation may be due to the Client being absent for a period due to holidays or illness, or because the Client wishes to appoint someone to assist him in his duties.

Although the contract is not specific, the notice should identify who the Client is delegating to, what their authority is and how long the delegation will last, so the Consultant is in no doubt as to who has authority under the contract.

It is important to recognise that in delegating, the Client is sharing an authority with the delegate who will then represent him, and he is not passing responsibility on to them.

The authority is shared, but ultimately, the responsibility will remain with the Client as the principal.

Note that there is no provision for the person with delegated authority to further delegate to others, the only delegation is by the Client.

It is also important to note that one can only delegate outwards from the principal, i.e. no-one can assume authority possibly because, within an organisation, they are senior to the Client.

The Adjudicator

The Adjudicator can be chosen by one of three methods:

1 By the Client specifically naming the individual in the Contract Data. The Consultant therefore knows at the time of tender who the Client has named and can therefore confirm that he does not agree with the Client's choice. This has advantages in that the parties always know who the Adjudicator will be if they have a dispute. The disadvantages are first that parties have commented in the past that naming the individual within the Contract Data assumes that there will be a dispute, and often the Adjudicator will require payment for having his name in the contract and being available should the parties have that dispute!
2 By the Client naming an Adjudicator Nominating Body in the Contract Data. This would normally be a professional institution. This tends to be the favoured method.
3 By the parties agreeing who will be the Adjudicator at the time of the dispute. The disadvantage of this method is that once the parties are in dispute, they probably would not readily agree with each other on anything.

The Adjudicator becomes involved only when either contracting party refers a dispute to him. As an impartial person, he is required to give a decision on the dispute within stated time limits. If either party does not accept his decision, they may refer the dispute to the tribunal (litigation or arbitration).

The Dispute Resolution Service Contract requires that payment of the Adjudicator's fees is shared by the parties unless otherwise stated (see also Chapter 10).

Differences between the Professional Service Short Contract and the Professional Service Contract

It is not intended to carry out a detailed "clause by clause" comparison between the clauses of the Professional Service Contract and the Professional Service Short Contract within this book; however, there are some important differences of a general nature.

- The Service Manager is absent (as stated above), but the Client has powers to delegate as necessary to suit his organisation and his particular purposes.
- There are no Main Options; instead, the flexibility and allocation of risk and payment is covered within the Price List.
- There are no Secondary Options, the reason for this is that these would probably not be used in the type of contract for which the Professional Service Short Contract is designed.

  However, delay damages and references to the Housing Grants, Construction and Regeneration Act 1996, which are both secondary options in the Professional Service Contract, have been included within the Contract Data for the Professional Service Short Contract.
- Some clauses which are unlikely to be used have been omitted, but if required can be incorporated by appropriate drafting of the Scope or the use of additional conditions of contract.
- There is no procedure for acceptance of Subcontractors, but responsibility of the Consultant for the performance of Subcontractors is clearly stated.
- In the Professional Service Contract, there is a prescribed list of items which must be included in the Consultant's programme.

  All of these may not be necessary for a project using the Professional Service Short Contract (if a programme is required at all), and therefore any detailed requirements for the programme within the Professional Service Short Contract are to be stated in the Scope.
- The payment procedure in the Professional Service Short Contract is different from that in the Professional Service Contract.

  In the Professional Service Short Contract, the Consultant is required to submit an invoice for payment after which the Client is obliged to pay after making any corrections he considers necessary. There is no payment certificate as in the Professional Service Contract.

  Also, the definition of Consultant's Defined Cost is much simplified. There is no Schedule of Cost Components; instead, cost is defined in terms of payments made by the Consultant for two items.

  i People employed directly or indirectly by the Consultant, which is calculated by multiplying each of the People Rates by the total time appropriate to that rate.
  ii Work subcontracted by the Consultant; the amounts paid to the subcontractor.

- The list of compensation events has been reduced from sixteen in the Professional Service Contract to eight in the Professional Service Short Contract, all of which are self-explanatory.

Assessment of compensation events is similar to that of the Professional Service Contract except that effectively, the rates in the Price List are used where possible.

Contract Strategy

The Professional Service Short Contract does not give the Client the choice of contract strategy in the form that is provided in the Professional Service Contract.

However, the Price List can be used to produce a lump sum contract or a bill of quantity-based contract. Target contracts and cost-reimbursable contracts are not provided for as they are regarded as unsuitable and too complex for this type of work. Similarly, Management Contracts will not be used on the type of work which the Professional Service Short Contract is designed for.

When the Client prepares the contract, he must decide the strategy he wishes to employ and hence who should carry the risks of quantities, pricing, etc.

The Price List

The Price List can be used in a number of different ways:

a as a lump sum contract based on the Scope including possibly a specification and/or drawings.

b as a lump sum contract where the Client gives a list of work items and the Consultant is required to price each item as a lump sum. The total of the lump sums is the lump sum offer for the service.

c as an admeasurement contract in which the Client lists quantities and item descriptions, and the tenderer enters a rate and extends the rate to the prices column.

Here, the itemisation is decided by the Client and is not subject to change except by compensation events. There is no method of measurement stated. If the actual quantities carried out are different from those stated, they are corrected.

d as an admeasurement contract in which the Client lists work items to be priced in the form of a lump sum and re-measurable items as in (c) above.

e any of the above (a) to (d) but where the Consultant writes the Price List to the instructions of the Client in the invitation to tender or as a result of negotiation.

Additional conditions

All changes to the core clauses should be included as "additional conditions" (NOT Z clauses) rather than amending the core clauses themselves; so, in effect, the clause remains in the contract, but is amended by the additional conditions. It is also critical that when drafting an additional condition, it must be clearly stated what happens to the original core clause, for example, it is deleted. If the original core clause is not deleted, then it is likely than an inconsistency will arise which will be interpreted against the party who wrote or amended the clause i.e. the Client.

These conditions may modify or add to the core clause to suit any risk allocation or other special requirements of the particular contract. However, changes should be kept to a minimum, consistent with the objective of using industry standard, impartially written contracts.

It must be remembered that if a Client amends a contract to allocate a risk to the Consultant that may have been intended to be held by the Client, the Consultant has a right to price it in terms of time and money; therefore, the practice of Clients amending contracts to pass risk to Consultants without considering who is best able to price, control and manage those risks can in many cases prove to be unwise and uneconomical.

It is important, particularly with the Professional Service Short Contract when the Contract should be kept "short" to spend time considering whether an additional condition is appropriate in each case, then when that decision has been made, that the clause is drafted correctly and aligned to the drafting principles of the original contract, in the case of the Professional Service Short Contract, useing ordinary language, present tense, short sentences and bullet pointing.

Some suggestions for additional conditions which are aligned to the Secondary Options from the Professional Service Contract and which can be adapted from them are listed below (note these have been discussed in earlier chapters, but to prevent the reader having to cross reference them, there is some duplication of what has already been discussed previously).

Price adjustment for inflation

The default within the Professional Service Short Contract is that the contract is "fixed price" in terms of inflation i.e. the Consultant has priced the work to include any inflation he may encounter during the period of carrying out the service and inflation is his risk.

The Client should make the decision at the time of preparing the tender documents as to whether inflation for the duration of the contract is to be:

- the Consultant's risk – in which case the contract does not need to be amended.
- the Client's risk – in which case, he should include some provision to adjust the Prices to take account of inflation. Changes to the Prices arising from inflation can then be dealt with as compensation events.

Changes in the law

Again, the default is that the contract is "fixed price" in terms of changes in the law i.e. the Consultant has priced the work to include any changes in the law he may encounter during the period of the contract, again it is the Consultant's risk.

Again, the Client should make the decision at the time of preparing the tender documents as to whether changes in the law for the duration of the contract are to be:

- the Consultant's risk: in which case, the contract does not need to be amended.
- the Client's risk: in which case, he should include some provision to adjust the Prices to take account of inflation. Changes to the Prices arising from changes in the law could then be dealt with as compensation events.

Parent company or ultimate holding company guarantee

This form of guarantee is given by a parent company (or holding company) to guarantee the proper performance of a contract by one of its subsidiaries (the Consultant), who, whilst having limited financial resources himself, may be owned by a larger and more financially sound parent company.

If a parent company guarantee is required, it should be provided by the date the contract comes into existence, or shortly afterwards.

The parent company guarantee should have an expiry date, which could be Completion of the project, the Defects Date or may even include the six, ten or twelve-year limitation period following Completion of the Works to cover any liability for potential latent defects, the expiry date being defined within the Contract Data

Parent company guarantees are normally used as an alternative to a performance bond.

Performance bond

A performance bond is an arrangement whereby the performance of a contracting party (the Principal) is backed by a third party (the Surety), which could be a bank, insurance company or other financial institution, that should the Principal fail in his obligations under the contract, normally due to the Principal's insolvency, the Surety will pay a pre-agreed sum of money to the other contracting party (the Beneficiary).

Whilst bonds are not always linked to insolvency, contracts normally give remedies in the event of default e.g. Defects provisions, retention, withholding of payment, delay damages, etc. It is where the defaulter is not able to provide a remedy as he is insolvent that bonds normally provide the remedy.

This bond may be between any two contracting parties, Client and Consultant, Client and Subconsultant, Client and Supplier, or any other contractual relationship.

If a performance bond is required from the Consultant, it should be provided by the Contract Date. The amount of the bond, usually 10% of the contract value, must be stated, and the form of the bond must be in the form stated in the Scope. In the event of the default by the Consultant, the Client is not paid the full amount of the bond, but the amount of the bond is the maximum amount available to meet the Client's costs incurred as a direct result of the default.

Sectional Completion

This is a provision that many have added into the Professional Service Short Contract, if the Client requires sections of the service to be completed by the Consultant before the whole of the service is completed.

Bonus for early Completion

Often, early Completion would be of benefit to the Client, for example, the Consultant is providing a service in respect of a sporting or cultural complex, and early Completion would allow a sporting event or concert to take place, bringing revenue to the Client.

Some consideration could be given for the Client to pay a bonus to the Consultant for early completion of the service.

Key Performance Indicators

In recent years, Key Performance Indicators (KPIs) have become a regular provision within construction contracts, the performance of the Consultant being monitored and measured against KPI's, with incentives set out in the contract for exceptional performance.

The Consultant normally reports his performance against the KPIs to the Client at intervals stated in the contract and is paid the amount stated in the contract if the target for a KPI is improved upon or achieved. Note that there is no payment due from the Consultant if he fails to achieve a stated target.

Typical KPI's

- Improved Client satisfaction
- Reduction of defects
- Savings in time
- Savings in cost
- Improved productivity
- Improved profitability
- Reduced health and safety incidents
- Reduced sickness absence
- Improved qualifications and skills
- Reduced impact on environment
- Whole life performance
- Reduced waste

It is critical in setting KPI's that the SMART principle applies:

- Specific: It has to be clear what each KPI actually measures. A KPI to measure how effectively people generally do their job will not be effective. It must cover a certain aspect of what they do e.g. specific deliverables.
- Measurable: Each KPI has to be measurable so that one can compare actual values with original baseline values
- Achievable: Each KPI has to be achievable, one cannot aim for a goal that no-one will ever obtain.
- Relevant: Each KPI must be relevant to the tasks in hand in order that improvement in delivering that task can be monitored and measured.
- Time-based: Each KPI must be time-based, ideally measured on a periodic basis.

Note that in January 2019, the publishers of the NEC4 Contracts issued amendments to the Contracts. Where appropriate, the amendments to the NEC4 Professional Service Short Contract have been incorporated within this chapter.

Index

For Product Safety Concerns and Information please contact our EU
representative GPSR@taylorandfrancis.com
Taylor & Francis Verlag GmbH, Kaufingerstraße 24, 80331 München, Germany

www.ingramcontent.com/pod-product-compliance
Ingram Content Group UK Ltd.
Pitfield, Milton Keynes, MK11 3LW, UK
UKHW050754180926
475701UK00007B/1037
* 9 7 8 0 3 6 7 8 9 6 0 2 7 *